THE *sustainable* HOMESTEAD

First Published in 2023 by Cool Springs Press, an imprint of The Quarto Group,
100 Cummings Center, Suite 265-D, Beverly, MA 01915, USA.
T (978) 282-9590 F (978) 283-2742 Quarto.com

27 26 25 24 23   1 2 3 4 5

ISBN: 978-0-7603-8048-2

Digital edition published in 2023
eISBN: 978-0-7603-8049-9

Library of Congress Cataloging-in-Publication Data

Names: Ferraro-Fanning, Angela, author.
Title: The sustainable homestead : create a thriving permaculture ecosystem with your garden, animals, and land / Angela Ferraro-Fanning.
Description: Beverly, MA : Cool Springs Press, [2023] | Includes index. | Summary: "Whether you're just dreaming, working on site selection, or an experienced homesteader, The Sustainable Homestead is the go-to resource to bring permaculture techniques to your crops, animals, and more"-- Provided by publisher.
Identifiers: LCCN 2022045245 (print) | LCCN 2022045246 (ebook) | ISBN 9780760380482 (trade paperback) | ISBN 9780760380499 (ebook)
Subjects: LCSH: Organic gardening. | Vegetable gardening. | Permaculture. | Animal culture. | Handbooks and manuals.
Classification: LCC SB453.5 .F47 2023 (print) | LCC SB453.5 (ebook) | DDC 635/.0484--dc23/eng/20220928
LC record available at https://lccn.loc.gov/2022045245
LC ebook record available at https://lccn.loc.gov/2022045246

Design: Cindy Samargia Laun
Page Layout: Cindy Samargia Laun
Photography: Angela Ferraro-Fanning; Darya Sedyh on pages 11, 14, 78, 122, 128, 135, 138, 166, 169, 170, 180, and 183
Illustration: Angela Ferraro-Fanning

Printed in China

CREATE A THRIVING PERMACULTURE ECOSYSTEM
WITH YOUR GARDEN, ANIMALS, AND LAND

THE *sustainable* HOMESTEAD

ANGELA FERRARO-FANNING FOUNDER OF AXE & ROOT HOMESTEAD

FOREWORD BY TEMPLE GRANDIN

CONTENTS

Foreword
BY TEMPLE GRANDIN

There is a great need to move toward more sustainable agricultural practices. For example, paying attention to processes such as crop rotation and good grazing practices can improve soil health. How do we get there? In many different industries, it is well known that it is the small operators who are often the most innovative. This is often true in agriculture, where small producers will start using a practice that the large producers may consider far out and fringe. A good example is in the broiler poultry industry. Fifteen years ago, feeding probiotics was considered fringe. Today, it is now mainstream for many large chicken producers.

Angela Ferraro-Fanning, in her book *The Sustainable Homestead*, shares the practices she has learned, and those that have helped her most, when growing her own food on a small piece of land. The recommendations in this book are based on more than a decade of experience combined with reading a variety of different sources such as university extension publications. This means she also learned from some of her failures. (She learned the hard way that young trees and livestock should never be mixed because the animals damage the trees.) Her book is recommended reading for individuals who want to avoid some of those mistakes and create a homestead where they do their own sustainable farming. In fact, there are some principles Angela shares that mainstream agriculture may have to adopt in the future.

Both Angela and myself are visual thinkers. When we think about how to solve a problem, we see pictures. Gradually a database of pictures is formed of things that work and things that do not. Visual thinking is a unique way of solving problems. (You may want to read my book *Visual Thinking*.) In any case, if you are also a visual thinker, you are sure to appreciate the carefully rendered infographics and the photos that relate to the text.

It is important to remember that while many of the recommendations in this book are general principles, others may only apply to the climatic conditions where Angela lives (in the northeastern United States). For example, Angela warns the reader that recommendations for keeping bees are very region-specific. For readers who live in other areas where the climate is different, it is always important to get good local advice. Still, I believe people who are considering growing their own food will find this book very helpful.

Introduction

Like many of my fellow self-taught farmers and homesteaders, I left the daily grind of a nine-to-five job for this life. I swapped sitting in front of a computer screen for more time spent outdoors, aligning my eating, daily practices, and lifestyle with the seasons and Mother Nature.

After a decade in graphic design, I was successfully self-employed with a full client roster. My work appeared in design magazines and annuals, and I won competitions. Wasn't I living the American Dream? It turned out that the dream simply wasn't mine. It took a struggle with postpartum depression for me to change my path. I needed to correct my course, even if it meant closing the business that I had worked so hard to build.

I talked about my desire for big changes with my husband, Shawn: What if I weaned myself off of my clients and income, grew as much of our own produce as possible, learned to can and preserve for future seasons, and drastically reduced our reliance on the grocery store? I had always had a small vegetable garden and a love of ornamental plants. This, by no means, was a stretch of the imagination for my husband. He supported me—in fact, he encouraged me. And for that I'm grateful.

My journey into homesteading began. Quickly our ¾-acre (0.3-ha) plot in a suburban New Jersey community began to morph into an urban farm. I replaced ornamental landscaping with edibles. I learned how to grow vertically

Upon purchasing the property, there was no horse stable. But there was a three-car garage, perfect for retrofitting into the stable we needed.

because space was limited, and I learned to forage in the nearby woodlands. I planted fruit trees and cherry bushes; lined our driveway with lavender, asparagus, and sweet potatoes; and learned the importance of companion planting because my neighborhood was heavily wooded and home to wildlife.

Experiencing the value that homegrown produce provided, my husband was the first to suggest the next step: Wouldn't it be nice to have fresh eggs? After a bit of research, I decided that ducks were a better fit for us than chickens. A few months later, a coop and three Cayuga ladies resided in our backyard. We were the talk of the neighborhood, and we had visitors over to see our garden and the new hens. Just six months after that, our first egg was laid. And the addiction grew.

I started researching tree tapping and discovered we were surrounded by sugar maples on the property. With ease I learned to tap trees and began making homemade maple syrup. I ordered a small plastic greenhouse and tried to keep edible vegetation alive as long as I could during the cold months. I remember cowering behind my countertop for fear of an explosion in my kitchen as I learned how to water-bath can and to make my first batches of soap. And then I dove headfirst into beekeeping.

The Three Pillars of Permaculture

1. Care for the Earth.
2. Care for the people.
3. Return surplus to the land or share with the people (fair share).

I became obsessed with farm-to-table and garden-to-table living. I read as many books as I could find and drank every drop of homesteading knowledge that I found. I was sleeping again, my depression was gone, I felt I had purpose, and I was feeling more connected to Mother Nature than I ever had. Eventually our ambitions outgrew our space. After a battle with neighbors over our Nigerian Dwarf goats, we resigned ourselves to finding a bigger and better homestead. It wasn't easy, but eventually we found the historic farm (founded in 1775) that would become Axe & Root Homestead. It had a barn, a three-car garage I could convert into a stable, space for pastures, a stream for our ducks to swim in, and outbuildings.

The new farm was my playground, and my largest stress was what to tackle first. The property had not been home to animals since 1950. At that time, it was a cattle farm and remnants of barbed wire and rotted fence posts could be found intertwined with barberry growth and Osage orange trees. There was so much work to be done. But there were century-old apple trees, a few rotting raised beds from an old garden, fruiting pear and cherries, a greenhouse, a coop, and a lovely farmhouse to call home. And it was ours.

The house and barns had been empty for well over a year, and the only tenants were mice, rats, ticks, spiders, and yellow jackets. I adopted two mousers to take care of the rodent population in the outbuildings. I thought it was so interesting that cats could be employed to tackle our pest issue. Could other animals do the same? What could remove the yellow jackets? Did anything eat ticks? Could I holistically bring this farm back to life? I started thinking of my approach as "holistic homesteading," but it turns out there was another name for it: permaculture. And down the rabbit hole I went.

I learned that the conventional agricultural practices I had thought of as essential for producing high yields and as a requirement for credible farming were having a negative impact on the environment. I became obsessed with the idea of growing food with complete respect for Mother Nature, incorporating animals and crops in symbiotic relationships. By day I worked outdoors, and by night I pored over websites, books, scientific studies, and articles, trying to absorb as much as I could about permaculture practices. The benefits of companion planting, perennial crop integration, composting, pasture rotation, cover crop implementation, plant and animal diversity, ecological design, food forest installation, capitalization of water, and soil health presented a new approach to farming that I hadn't known existed. And soon I learned the powerful result of all of these efforts combined: land regeneration.

The very idea that I can grow food and sustain animals and wildlife in a fashion that actually helps to improve the soil is still fascinating and inspiring. Everything we need to successfully grow food and feed ourselves is already provided by way of sun and rain, soil, microbes and beneficial bacteria, carbon and nitrogen from the atmosphere, and the natural decomposition process that produces nutrient-rich humus. After all, this is the way nature intended.

The journey to a holistic, permaculture, sustainable farm is an ever-evolving one, and I am learning constantly, adapting as Mother Nature asks me to. It took me years to learn these concepts and I wish I had had a cohesive resource to draw information from—one specifically for the homesteader. It is my hope that *The Sustainable Homestead* will be that resource for you.

The Twelve Principles of Permaculture

1. Observe and interact.
2. Catch and store energy.
3. Obtain a yield.
4. Apply self-regulation and accept feedback.
5. Use and value renewable resources and services.
6. Reduce waste.
7. Design from patterns to details.
8. Integrate rather than segregate.
9. Use small and slow solutions.
10. Use and value diversity.
11. Use edges and work with margins.
12. Creatively use and respond to change.

Following the permaculture principle of "using the margin." Sheep graze outside the pasture within a portable electro-net fence while pasture grass regrows.

CHAPTER **one**

SITE ASSESSMENT

No plot of land is truly tenant free. Even if there is no house, no barn, and no human living there, it's already home to a myriad of plants and wildlife. Humans often are guilty of thinking that a new acreage is a clean slate, or that an abandoned or temporarily vacated farm is uninhabited. But it isn't really. An ecosystem is in place, from the ants crawling below to the honey bees flying overhead. There could be a stream bubbling nearby or a local fox in her den. A highly sophisticated network of mycorrhizae already is developed underfoot, and billions of bacteria, protozoa, and microbes have been in existence since long before. These elements are contributors to the greater whole. They should be encouraged to thrive and coexist alongside a permaculture homestead.

Rather than play the role of the homestead architect immediately, there is a clear benefit to watching and waiting in order to learn the nuances of a property. Observe how the sun rises and sets and where the wind tends to whip. How does water flow and pool during extreme rainfall? Where does the snow drift? The plants and animals that already live on the land know the answers to these questions. For example, burdock grows in poor, infertile soil. Plantain prefers soils with little water retention. Wild violets thrive in damp, shaded locations. If heron fly overhead, there is a natural waterway nearby, most likely home to minnows or small fish. A prevalence of opossum most likely indicates a high tick population. It is up to the homesteader to learn these patterns and embrace them as communications from nature.

« This flood plain is home to many naturally growing grasses and forage. In good weather conditions, the plain is used as an added pasture space.

STEP 1: EVALUATE YOUR SITE

Observe and Consider the Elements

There is a spot on my land that is perfectly flat. The soil is fertile and soft, lush with green grass and weeds. The trees line the perimeter of the field, and it receives full sun. This is the perfect area for a garden, or so I thought until I observed the site for one full year. Turns out, this area is nutrient-rich because it's where all the water from the homestead pools. In heavy rainfall it floods altogether, and the soft pasture space is host to waves and rushing water several feet high. Debris including fallen trees, neighbors' trash cans, lawn chairs, and fence posts all rush through this highway of water in a heavy storm. While the flooding does recede a short time later, the waterlogged footing would drown the roots of any crop that requires well-draining soil. Imagine the disaster that would have ensued if I had installed a fence, raised beds, pathways, and crops. It is no site for a garden.

Instead, I began reconstructing the garden space that already had been worked by previous homeowners. It was heavily riddled with weeds, barberry, snakes, and insects; a thriving ecosystem. A pear tree resides on the northeast corner, and I thought it silly a raised bed had been installed underneath. Turns out, sunlight can filter through bare branches before leaves sprout providing direct sunlight to early spring crops. The same tree, after foliage appears, can host shade to crops below that benefit from shelter from the harsh summer sun, such as lettuces and even brassicas. There's so much valuable insight that can be gained from experiencing the shift in seasons, weather, and temperature before making any changes to the footprint of the property.

Effective rainfall is the amount of rain that is absorbed and retained by a landscape. Ineffective rainfall is the water runoff, or the water that is not absorbed. Soil rich in structure

This site is perfect for a market garden. Luckily, with time and observation, it was discovered to be a flood plain.

and fertility will have much effective rainfall. It has the ability to withhold water and survive longer periods of time without irrigation. I learned quickly that because my property resides on a hillside and much of the earth was clay, I had very little effective rainfall. We receive plenty of rain in my corner of New Jersey, but my landscape just wasn't retaining it. All of these observations came together to shape the course of action I would take in finding the right locations for plants, installing raised beds or other growing spaces, and even determining where I introduced fencing and posts.

When assessing your property be sure to note where water pools, how quickly it absorbs, and how the landscape shapes the flow of water. If snow is present during portions of the year, does it drift? Do straight-line winds affect portions of the property in any way? What areas of the real estate receive full sun, partial sun, and shade? Where does the sun rise and set with relation to the land? These considerations are just as pivotal to the success and efficiency of the homestead as the footing, soil, and terrain itself.

Observe and Consider Wildlife

One of the first leaps homesteaders make into animal ownership is through poultry. Chickens, ducks, geese, guineas, quail, and turkeys all contribute to the farm in many ways besides providing fresh eggs (detailed in chapter 4). Unfortunately, many predators favor these animals as prey. If the intention is to keep poultry, rabbits, barn cats, or other small barnyard family members on the homestead, are there any known predators in the area? And if so, where are their high traffic or migration routes? How will you keep your animals safe?

When considering predators, use your common sense. By this, I mean do not place a chicken coop or run, for example, within the direct path of foxes or coyotes. Here's another example: In bear country, beekeepers can certainly keep hives, but to ensure safety, the hives may need ratchet strapping and even electric fencing if bear activity is high. Even in New Jersey I had my own hives ransacked and broken by black bear before I learned to ratchet strap them together. Talking to locals within a close distance to your property can be an excellent resource if they are willing to share their experience and their deterrents for local wildlife.

In addition to site selection, it is imperative to evaluate any local predators and wildlife with regard to fencing. What type of fencing will need to be installed to farm alongside these predators? Barbed wire is illegal in many states as it causes injury to both intruders and livestock. Wooden fencing, welded wire, or hardware cloth can successfully deter unwanted bears, foxes, and more when accompanied by electrified wire. Hardware cloth measuring no more than ¼ inch (6 mm) is an effective enclosure for poultry and small hoofstock, and it should be buried down into the soil approximately 10 inches (26 cm) deep. Even better is to bend the hardware cloth outward, away from the coop or pen, at a 90-degree angle when burying within the soil. This will discourage any predators who may be determined to dig beneath the wall including fisher cats, weasels, and fox. If predator pressure is high enough, a Livestock Guardian Dog is a possible solution to accompany strong fencing systems (see page 106).

The local fauna wouldn't exist without flora, as brushes, brambles, grasses, and woodlands provide a food and shelter source for these creatures. A question worth asking before starting your homestead at any new location is, How does Mother Nature provide for these animals already? An oak tree that may drop thousands of acorns could be viewed by some

Phenology

One of the most fascinating topics I've come to learn about is phenology; the practice of reading patterns or relationships within nature when it comes to weather, animals, and vegetation. These same indicators are what farmers used as natural cues for planting timelines long before there was access to a weatherman. While some folks may view this as merely folklore, I have found that generally these planting guidelines are fairly accurate:

- When crocuses bloom, plant radishes, peas, kale, and chard.
- When daffodils, peaches, and plums bloom, plant carrots, beets, brassicas, and second radish crop from seed.
- When Oregon grapes blossom, transplant out brassica seedlings.
- When forsythia blooms, plant out onions.
- When dandelions open, plant out potatoes.
- When maple tree leaves leaf out, plant out new perennials.
- When lilacs and apple trees bloom, plant cucumber, bush beans, and squash.
- Right before lilacs wilt, plant out annuals.
- When lily of the valley bloom, transplant tomatoes.
- When apple blossoms fall, plant out corn, pole beans, basil, and marigolds.
- When iris and peonies are in full bloom, plant out peppers, eggplant, and melon.
- When morning glories bloom, Japanese beetles are arriving.
- A prolific berry season means a harsh winter ahead.

as a nuisance. But to surrounding wildlife, it's a pivotal source of food for successful hibernation. To farm alongside nature, instead of intersecting with it, it's important to work with these potential obstacles and find compromises. For example, the same acorn tree could provide forage for pigs through pannage (the practice of finishing pigs or hogs on fallen acorns before processing as explained in chapter 4).

Any existing vegetation should become a factor in what animals you may choose to bring into your homestead and what foliage is available for them to forage. I had considered introducing goats to the new farm, but when I found we have much more leafy and grassy growth than brambly, fibrous shoots, I opted for sheep. The landscape and the forage it provides can guide your path.

STEP 2: DEVELOP A HOMESTEAD LAYOUT

Homesteading is challenging. The work can be less so when a proper homestead layout is implemented to make hauling feed, moving animals, conducting chores, dumping compost, and providing water less physically taxing. Before creating your homestead site plan, there are a few points to consider in order to make sure your plot is as efficient and effective as possible. As mentioned in step 1, be sure to observe and gather as much information as possible about your site before installing any infrastructure. A full year of data collection is ideal to ensure you fully understand the effects each season brings.

Once you have a rough idea of where you want gardens and livestock, then you can make more informed choices about your investments. For example, placing pastures adjacent to animals makes sense, rather than having to lead animals far distances from point A to point B. Ensuring water spigots are installed where irrigation may be needed for crops is better than connecting long hoses. Some considerations may conflict with information you've gathered. For example, should you want to eat from your kitchen garden daily, perhaps placing it on the far end of your property, well away from your home, isn't ideal even if that's where the site is best. (Perhaps you could consider an additional smaller garden placed close to the kitchen for your most-used crops?) Other fundamentals to consider are included on the following pages.

Land and Elevation Changes

Hills and elevation changes can result in land erosion and water runoff. Growing crops as well as housing or grazing animals can be challenging if there is a constant or steep incline. After assessing the landscape, consider ways to incorporate growing and grazing spaces that retain water and follow the natural terrain. Bioswales can be created and installed by mounding soil and organic matter parallel to a slope. This allows any water runoff and nutrients it may be carrying to be captured and held within the swale (see page 58). A hillside of bioswales is an effective way to grow crops in less-than-ideal terrain. Those same hills can be difficult for the homesteader to climb, especially several times per day conducting chores and tending to animals.

Weather can also affect how you and your animals can traverse different elevations. If winter weather is a factor, how will ice and snow affect traveling by foot (or hoof)? Ensuring that stairways or gentle grades are installed to create safe points of passage is essential for both farmer and any animals traversing the slope.

When designing the layout for the homestead, consider livestock and their routes from barn or stable to pasture.

Guardian geese make excellent watchdogs for other flock members while swimming or free ranging.

Ducks and geese bathe in the unfrozen stream in mid-winter.

A natural waterway on the property is perfect for keeping ducks and geese. This reduces the amount of water that would otherwise be used filling pools for bathing.

Water and Irrigation

Irrigation is pivotal for growing most fruit and vegetable crops. While growing native plants and forage can heavily reduce the amount of water needed to sustain plant life, chances are that a homesteader will need multiple water spigots for hydrating animals and crops. Be sure to think through how water will make its way to the garden and animal pasture and housing spaces. For example, if working within a desert climate, are waterfowl and aquatic life easily incorporated into your space? A homestead with natural waterways could provide ducks and geese with a place to swim without requiring additional daily water usage to keep them happy and provide lots of water to drink.

Just as important, be sure to consider natural water accumulation. Are there any low-lying wetlands that could cause hoof rot and damage to hoofstock if used as pasture? Does a hill encourage rain runoff right into your barn? If drought is inevitable in your location, perhaps an easily accessible water catchment pond could be installed along with rain barrels for both animals and crops (more on this starting on page 58).

Water will naturally flow from higher points on the land to lower points, such as valleys and ditches. Even relatively flat properties possess small ridges or slight grades where water takes a course from higher to lower elevation. Understanding the course that water flows throughout the landscape can help the homesteader to determine where to place ponds, swales, and other forms of water catchment systems to capitalize on rain and groundwater (see page 58).

A topographical map is helpful in identifying these shifts in elevation; the general contour of the landscape and the flow of water can be better understood by looking for key lines. A key line is the line that resides between elevation changes on the map; in other words, where the lines begin to gain space between them. Following the contour and shape of the land with respect to key lines benefits plowing, seed sowing, and pasture placement by ensuring optimal water exposure, absorption, and/or flow.

Create a Permaculture Zone Map

One of my favorite permaculture concepts is the creation of a zone map. A zone or site map is a simple plan of your homestead divided into areas based on usage frequency. For example, on the map, zone 0 would be your home as it's the most frequented space on your farm. It's where you eat, sleep, and relax. Zone 1 would be areas you visit most often. This might be the barn you visit multiple times per day to tend to animals. Zone 2 might be your pasture, garden, and other growing spaces. You might visit these locations once per day or even less than that; maybe once every other day and so on. Keep dividing the map further and further based on how often you visit these spaces. In an ideal ecologically focused farm setting, the map would even include a zone for wildlife in an area that isn't accessed often by humans.

What this map accomplishes is a visual representation of how to arrange items based on efficacy and ease of access. Going back to the kitchen garden example, a kitchen garden quite literally over a river and through a set of woodlands wouldn't be very accessible from your home. By grouping items together, farm chores can be accomplished more efficiently. Encroachment on wildlife areas is reduced to an absolute minimum.

PERMACULTURE ZONE SAMPLE MAP

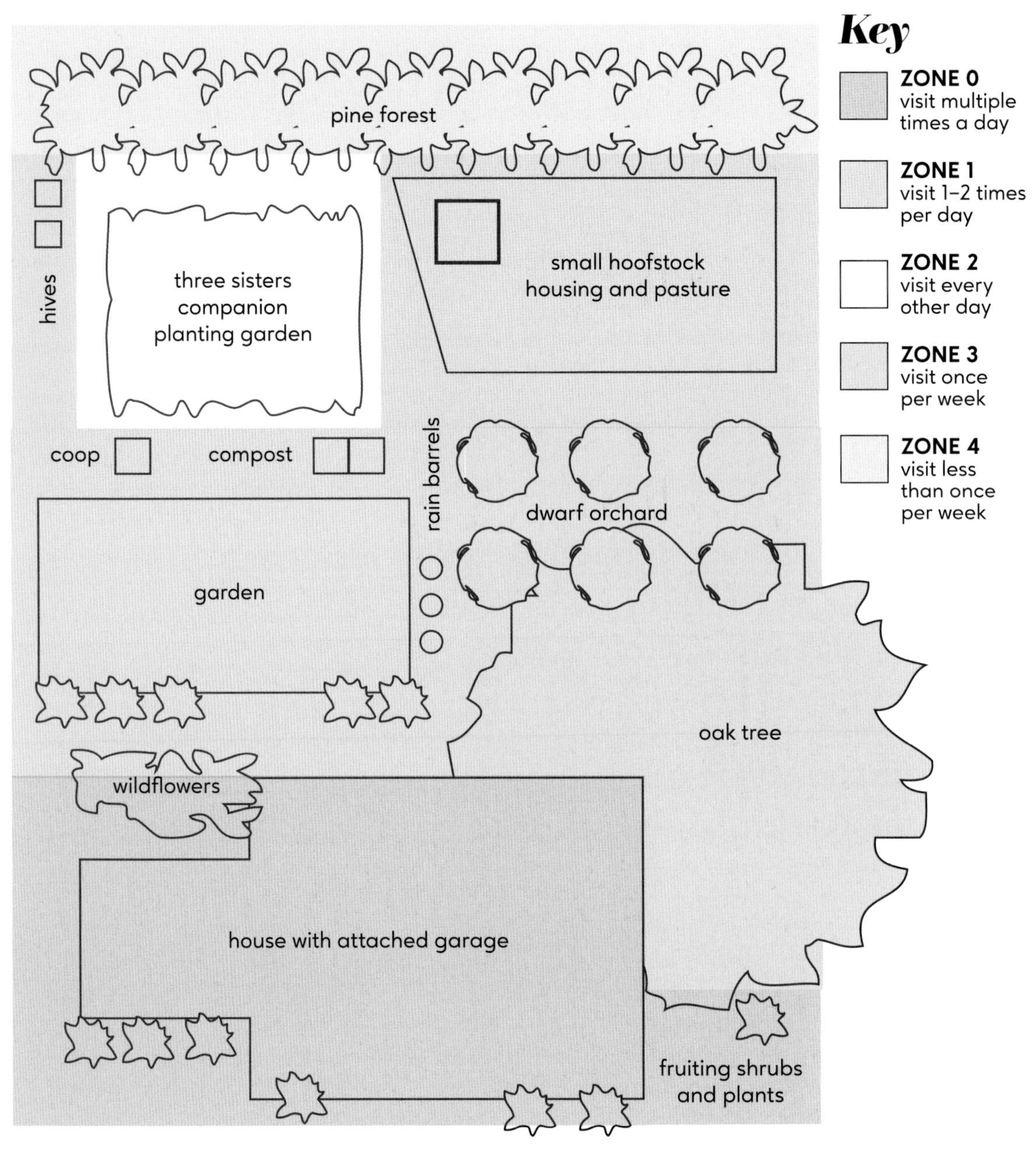

This map illustrates all of the elements included on a sample quarter-acre (0.1-ha) homestead. Once all items are accounted for, the map is divided into zones. Zones with lower numbers indicate areas that are visited most often. Higher numbered regions are those with less daily traffic. By grouping objects together based on the frequency in which they are accessed, a more efficient homestead design can be achieved. This also allows some portions of the property to be more inviting to local wildlife as they host less daily traffic.

QUARTER-ACRE (0.1-HA) SUSTAINABLE HOMESTEAD

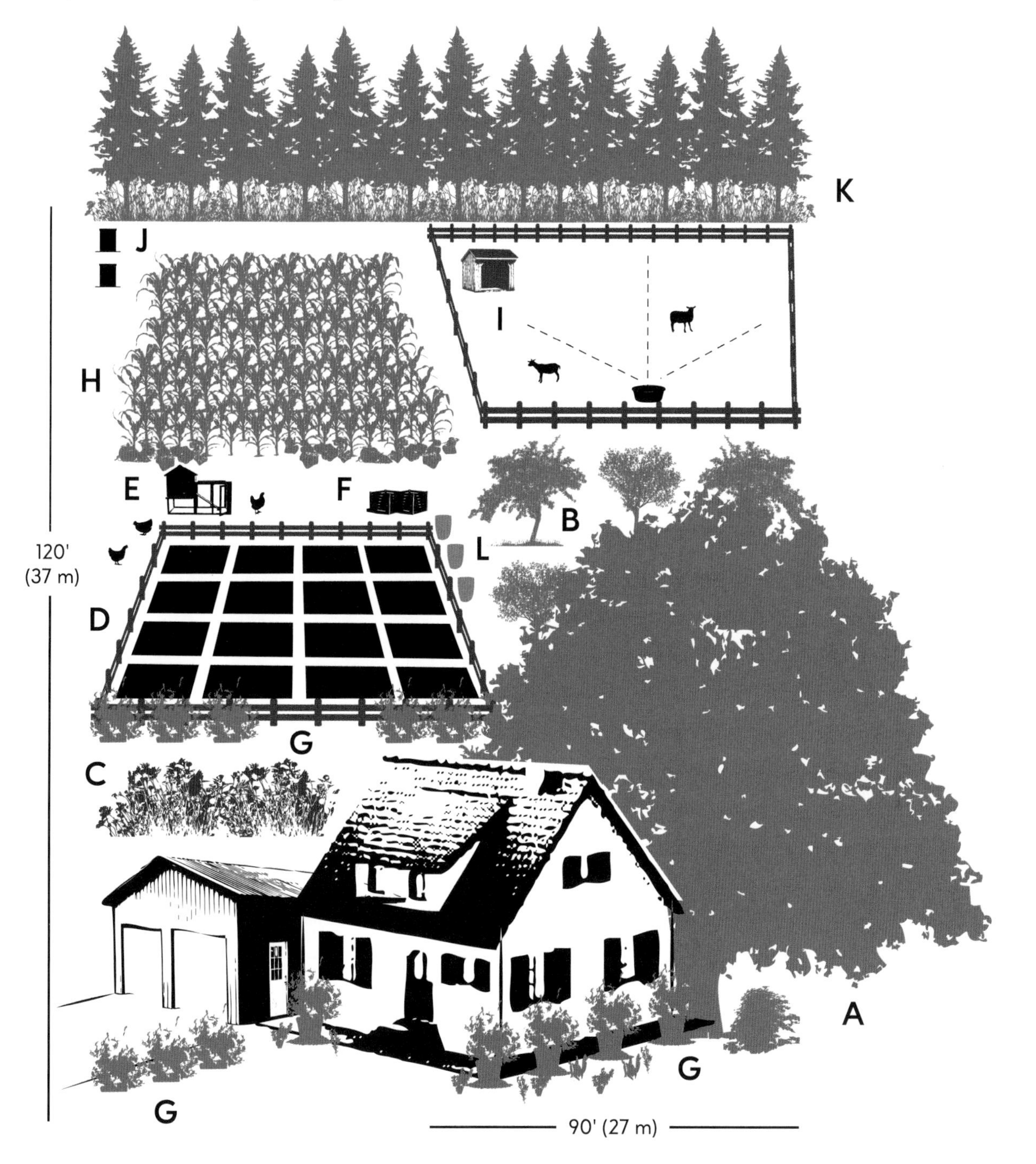

A: oak tree (overstory) / **B:** dwarf fruiting trees (midstory) / **C:** wildflower pollinator field / **D:** vegetable garden with 16–4' × 8' (1.2 × 2.4 m) raised beds / **E:** chicken coop / **F:** 2-bin compost system / **G:** edible bushes, flowers, and herbs / **H:** three sisters or companion garden / **I:** small hoofstock rotationally grazed / **J:** beehives / **K:** preserved woodlands / **L:** water catchment system (rain barrels)

HALF-ACRE (0.2-HA) SUSTAINABLE HOMESTEAD

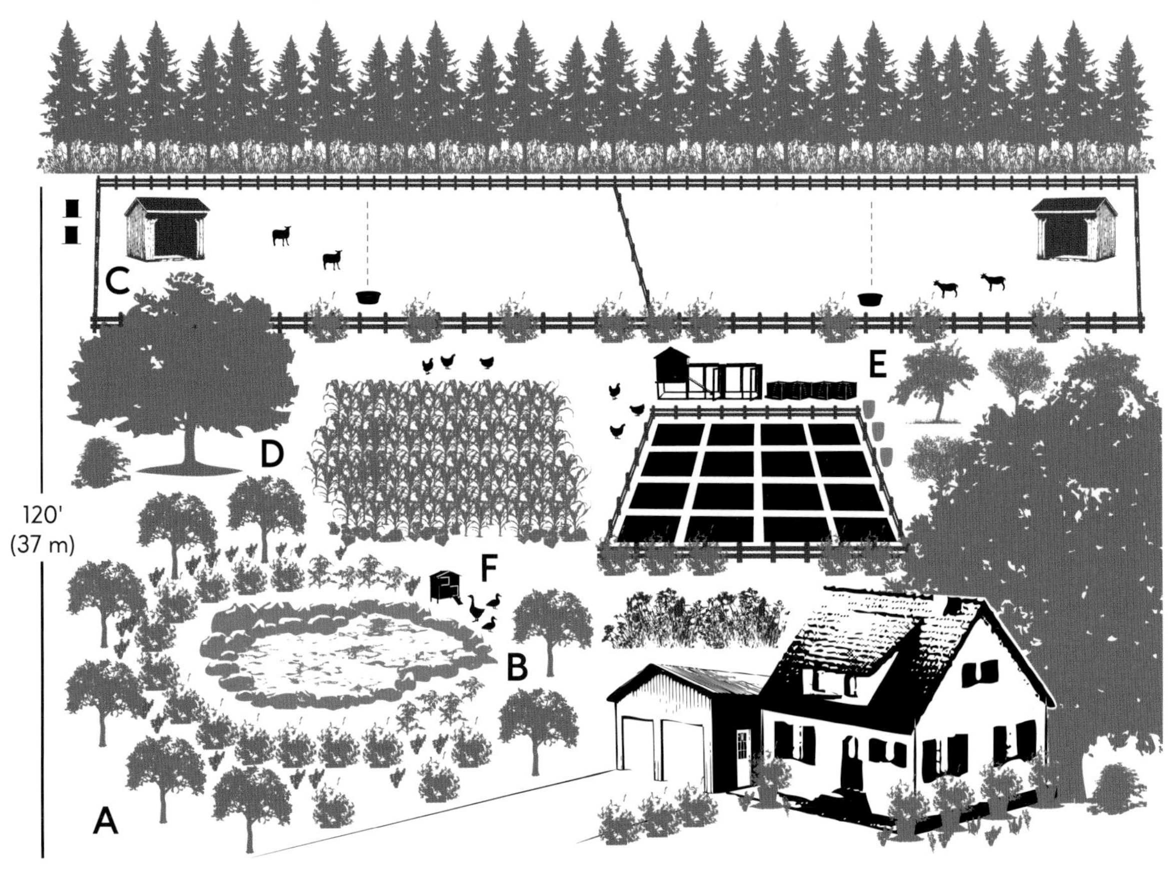

▲ **A:** mandala-style garden / **B:** catchment pond / **C:** additional pasture for two sheep or two goats, grazed rotationally / **D:** maple tree for tapping (overstory) / **E:** additional compost bins and rain barrels / **F:** ducks and geese with a small coop

This sample half-acre (0.2-ha) sustainable homestead includes all of the elements of the quarter-acre (0.1 ha) version. With the addition of another half-acre (0.2 ha), more flexibility is offered for additional crop growing spaces, a small catchment pond for ducks and geese to swim in, and even a second pasture space for the housing of small hoofstock.

▶ A one-acre (0.4 ha) plot can offer lots of real estate for crops, a food forest, animals, water catchment systems and even orchard trees. In this sample layout, the quarter- and half-acre (0.4 and 0.2 ha) homestead design has been expanded with the addition of more pasture space for small hoofstock. More pasture fencing offers opportunity to install fence-line swales where crops can be grown alongside the perimeter. This allows for water runoff from the pasture to deposit nutrients into the soil which is host to crops (see more information in chapter 3). More pasture space for animals also means more forage and, ultimately, more carbon absorption by way of regrowth.

ONE-ACRE (0.4-HA) SUSTAINABLE HOMESTEAD

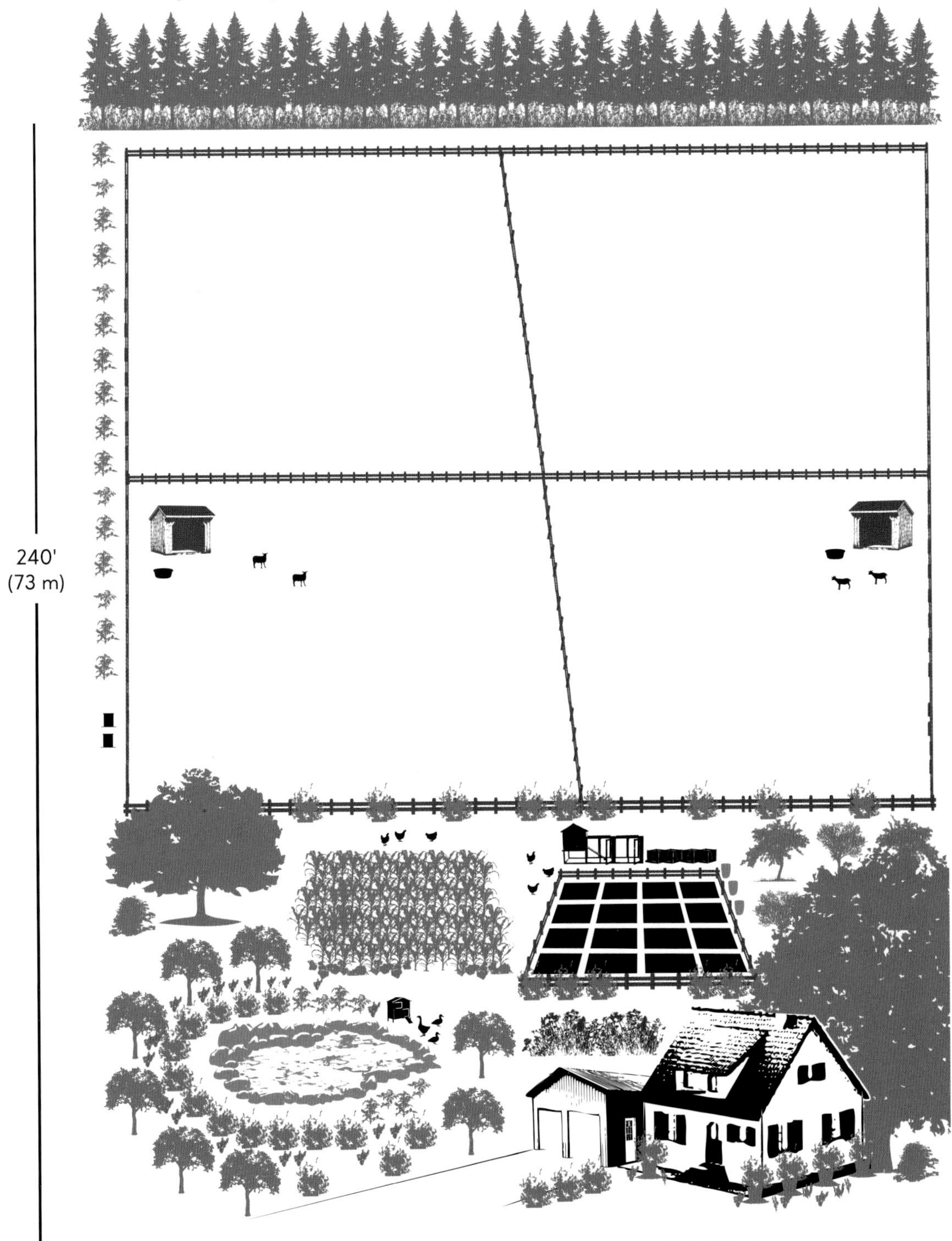

FIVE-ACRE (2-HA) SUSTAINABLE HOMESTEAD

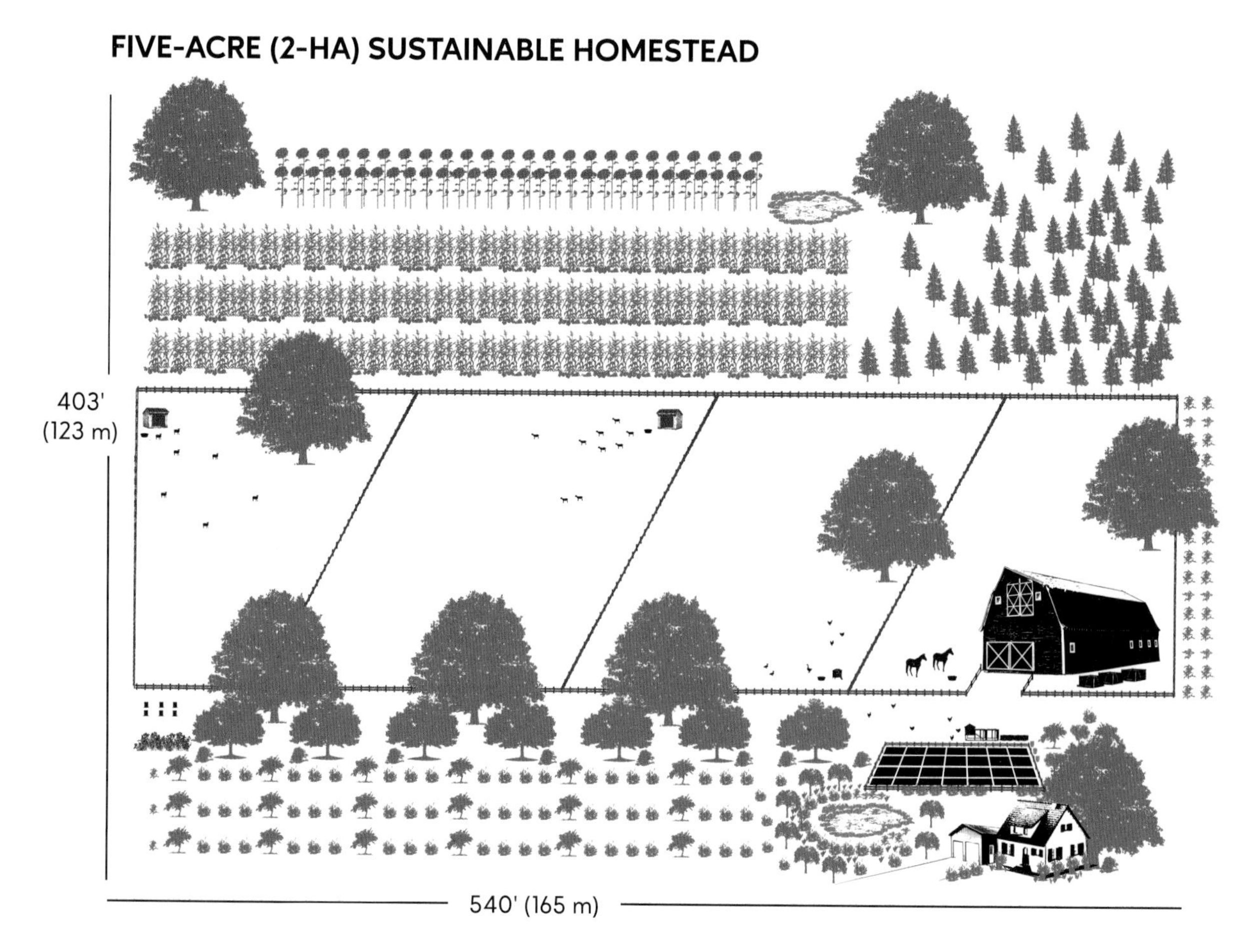

So long as terrain and soil types allow, there are endless possibilities for cultivating a sustainable homestead on five acres. The sample one-acre (0.4-ha) plot has simply been expanded with enough real estate for a barn, ample pastures for rotational animal grazing, a food forest complete with an overstory, fruiting trees, and understory plants. Multiple water catchment systems offer a reliable irrigation source for crops. A larger kitchen garden remains next to the house for ease of access, while livestock-appropriate compost systems have been installed next to the barn. Along with more room for homesteading activities, more land has been designated for preserved wildlife spaces.

STEP 3: UNDERSTANDING THE FOUNDATION

At this stage, most of the data has been gathered by way of simple observation. There is one more step of due diligence, however, that I highly recommend before establishing growing spaces or exploring cover crop and animal forage options. Most gardeners are familiar with the concept of testing their soil at the beginning and/or end of every growing season. A simple soil analysis is conducted to determine what nutrients remain within the soil and what items may need to be added for plant health and high yields. If no nitrogen is needed, and a well-meaning grower doesn't realize (and as a result adds an excess to the soil), plants will suffer and could even die off. Knowing what the soil requires helps to create a clear plan for adding amendments, determining which crops to grow where, and what cover crops may be needed, if any.

The same idea applies to the overall homestead, especially if a farm does not preexist on the property. Test the soil to understand what's available and what isn't in terms of nutrients. Test in multiple locations throughout the property to understand how these levels change. Explore what rock and mineral types are present. Is the earth primarily sand, silt, or clay? Are there soil contaminates that need to be removed and how can they be remedied? What weeds and native plants grow in future pasture spaces and, if they are poisonous to livestock, how can they be permanently removed and prevented from returning? Truly getting to know and understand the landscape, the natural ecosystem, and the access available to water and sunlight is essential to setting up the sustainable homestead for success.

Soil Type Identification Test

It's easy to determine the contents of your soil when it comes to sand, silt, and clay. A well-balanced soil containing roughly equal parts of sand, silt, and clay creates loam and provides vegetation with a fertile, ideal growing environment. More about loam and its contents can be found starting on page 40.

MATERIALS

- Garden soil
- Trowel (optional)
- 1 pint-sized (473-ml) canning jar with ring and lid
- Water

DIRECTIONS

Collect a handful of garden soil, either with your hand or a garden trowel. You can collect topsoil from the surface of your garden or dig down to 8 inches (20 cm) deep for a rhizosphere sample (more on soil layers in chapter 2).

1. Fill the jar approximately halfway full with garden soil.
2. Fill the remaining space within the jar with water, leaving 1 inch (2.5 cm) of headspace at the top.
3. Close the jar tightly. Shake the jar vigorously to suspend all of the particles within the water.
4. Allow the jar to rest in a level location for about four hours, or until all of the particles have settled.

After the sample rests, it is easy to see the layers within the jar. The bottom layer is sand and larger components, such as rock. The second-to-bottom layer is silt. The third layer from the bottom of the jar is clay contents. Finally, any water not intermixed with soil particles floats at the top.

This test communicates the approximate percentage of sand, silt, and clay within a given growing location.

CHAPTER *two* // SOIL

I still remember the day I slammed my shovel into the soil, breaking ground for the first thing I wanted to plant on my new homestead. I'd been preparing for this moment for months by carefully tending to a Blue Damson plum sapling in a large pot. Yet when the time finally arrived to dig a permanent hole for the tree at my new homestead, I was met with nothing but resistance. My shovel would just not break the ground. I set the shovel upright and jumped on top, trying to balance my weight and drive the shovel downward simultaneously. The shovel was bending—the soil was not. I discovered the soil wasn't really soil at all; it was just hardened earth devoid of any nutrients and water. It had been baked by the hot summer sun into a hard clay.

Despite the fact that this farm was host to cattle only sixty years before me, it had been treated with herbicides since that time. No crops had been grown; no animals had grazed; and the land I was trying to dig had not been allowed to grow fallow. Instead, the exact spot I had assigned to my new home orchard had been treated as a lawn space: a carpet of conventionally seeded grass, crab grass, and dandelions remained. At that time, I did not know that the plants that grew underneath my feet and spade were communicating the soil's status. Since then, I have come to learn that dandelions indicate soil compaction and a lack of calcium. Crabgrass tends to thrive in challenging soil conditions where there is less water retention, or too much.

Two hours later, the sapling was finally planted, covered with dirt and liberally watered. I had used a pickaxe and shovel to dig out the hole. While I dug, I had plenty of time to notice that only about 25 yards (23 m) to my left a trickle of water was coming from the top of the hillside. The soil immediately surrounding it was thriving with lush green grass and moist black soil. A giant oak tree shaded the area and underneath grew a fruit tree I'd not yet identified adjacent to an heirloom pear. A mess of wild blackberry and wineberry canes rooted themselves among fallen oak leaves that formed a ground mulch. This was peppered with acorns, orchard grass, and even a rogue pumpkin vine from the previous owners' compost heap. Milkweed speckled the outermost perimeter of this area and thrived in the unshaded borders. This was my first brush with permaculture and its relationship to soil.

« Finished compost should be thought of as a soil conditioner, rather than a fertilizer, as the nitrogen and other nutrients it contains are released slowly over time. Beneficial microbes, nematodes, protozoa, and more are ready to help build strong soil structure right away.

THE IMPORTANCE OF HEALTHY SOIL ON THE HOMESTEAD AND HOW WE GOT HERE

Healthy, nutrient-dense soil supplies plants with the elements that they need to create maximum yields boasting nutrient-dense produce and/or animal fodder. A plant's life cycle requires a total of seventeen elements, the majority of which are obtained from the soil. Plants source carbon, hydrogen, and oxygen from water and air; the remaining nutrients are provided by animal manures and soil amendments, such as fertilizer.

Thriving soil is rich in organic matter, nutrients, moisture, and microbes. These elements are the product of an entire functioning underground ecosystem of nematodes, fungi, mycorrhizae, earthworms, protozoa, minerals, decomposed and undecomposed plant and organic matter, and root exodates. These fundamentals are achieved through a balance of animal grazing, animal manure, perennial root systems, minimal soil disturbance, decomposition, soil protection by way of natural mulches, and plant diversity. All are necessary and required to feed and facilitate healthy soil by Mother Nature's design. When the soil's top and sublayers are properly nourished, so are the plants and crops that the soil produces, as are the animals and people who ultimately consume from it.

Over-Farming

Many folks in the agricultural world have heard the ideology that the majority of farmland today no longer holds the nutrients that it did in generations past *as a result of over-farming*. A loss of nutrients in the soil directly effects the nutritional density of the crops it produces. Less nutrient-dense food means less nutrition for the people consuming those foods. This is a problem.

Why It Matters

Studies have shown that there is a dramatic decline in the amount of nutrients in our food today compared with produce crops from several generations ago. Scientists attribute this reduction in vitamins and minerals of fruit and vegetables to crop manipulation. In other words, crop species cultivated today are selected for rapid growth and pest resistance. The trade-off is that these plants absorb fewer nutrients from the soil in exchange for larger, faster yields. Additionally, the soil is not replenished and, when combined with modern-day agricultural practices, soil health suffers.

Think about it: tilling land and exposing soil allows carbon, water, and key nutrients to leak into the atmosphere over time via evaporation. Nutrients and beneficial bacteria that may be overturned during the tilling process and are left to reside on the top of exposed soil are solarized and lost. Root systems and plant matter that were once vessels for absorbed carbon are uprooted and unprotected by the soil. The carbon dioxide that was once stored is rereleased into the atmosphere.

Before homesteading and agriculture was ever on my personal radar, I thought the term *over-farming* implied that the land has provided too many harvests and was exhausted of some form of energy. Perhaps, I thought, there was a maximum number of yields any individual piece of land could offer. Now I realize that, in essence, this basic understanding was somewhat correct. If any individual piece of land is not renewed by way of added organic matter, replenished nutrient loss, and polyculture, there is a maximum number of yields of nutritious food that a piece of land can provide.

The Underground Network

Let's go a step deeper and look at the roots. The roots of plants, be they annuals or perennials, are surrounded by microbes and mycorrhizae. Microbes essentially come together to coat the roots of plants creating a protective barrier from harmful pathogens, a concept called *induced systemic resistance*. Microbes also help to stimulate plant growth by releasing hormones. In exchange for their assistance, sloughed off plant cells, sugars, starches, and even amino acids from the plant's roots feed these microbes.

Mycorrhiza is a beneficial root fungus that grows in a symbiotic relationship alongside plant roots. The roots uptake nutrients and offer plant sugars that are essential to the health and survival of the fungi. In return, the mycorrhizae act as an extension of the plants' roots, offering hydration and nutrients for the roots to absorb. Mycorrhizae interact with the plants' roots as receptors providing key nutrients, such as phosphorous, when needed. Where mycorrhizae are present and available to assist, plants are additionally less prone to water stress.

All of this is good! The mycorrhizae and related systems do not contribute to the issues I am talking about with over-farming. Unfortunately, the delicate relationship between the roots of plants and surrounding microbes and mycorrhizae are severely damaged when synthetics are applied. Herbicides and pesticides are not able to distinguish between what is a weed, unwanted plant material, a pesky insect, and what is a helpful microbe or mycorrhizae. The working exchange system between roots and the surrounding fungi is degraded when manufactured fertilizers are applied as the plants ingest those synthetic nutrients from the application, and not from the surrounding soil and helpful fungi. As a result, these microbes and mycorrhizae are no longer needed and, thus, no longer fed. These beneficial organisms begin to die. Thus, the application of pesticides, herbicides, and synthetic fertilizers contribute to the degradation of healthy soil.

Decreased Diversity

Monocropping, or the practice of sowing and growing one single crop variety in a given space, is another detrimental act to soil health. One single plant species is not just a breeding ground for relevant pests and disease that affect those hosts—it's an army of plants striving to absorb the same nutrient panel from the soil. It does not matter if the monocrop is grass for animal grazing or a vegetable for human consumption. Single plant species cultivation results in nutrient loss. For example, a field full of corn is hungry for nitrogen and, if the nitrogen is not replaced, the soil will become depleted of this essential nutrient very quickly.

When synthetic fertilizers are added to compensate for nutrient loss, a large portion may never even see the soil as the fertilizers are prone to evaporation. Studies have shown that up to 40 percent of nitrogen urea fertilizer can be lost. A net loss of nitrogen is a result within the soil; a loss in dollars is a result for the farmer, too. A loss of nutrition in the soil directly translates to a loss of nutrition in our food and in our animals' forage.

SYMPTOMS OF POOR SOIL

It doesn't matter if you are gardening in a raised bed or farming a large plot of land, plants are usually the first indicators of soil trouble. When I became serious about growing as much of my own food as possible, I ran into issues that many gardeners are familiar with, such as yellowing leaves, blossom end rot, and wilt. I quickly learned that nutrient availability within the soil directly affected the overall health and well-being of my plants . . . and their visual appearance. Those yellowing leaves on my broccoli indicated a nitrogen deficiency. Blossom end rot on my tomatoes was a result of too little calcium within the fruit. A lack of copper was expressing itself as wilting in the leaves at the top of my pepper crops. Luckily, these problems were easily remedied in my raised garden beds with some simple soil amendments.

Sometimes, however, poor soil conditions make themselves known through more dramatic signs. Water retention and soggy ground indicate a lack of drainage within the soil. Where the earth lacks porosity and is completely saturated, algae and water-loving weeds will begin to accumulate. Conversely, soil that is not able to absorb water and allows rainfall or irrigation to run off can cause flooding and pooling in low-lying areas. This water runoff, especially during extreme weather events, is a sign of highly compacted soil or clay.

Without the ability to absorb water, subsoil layers can remain dry even after heavy rainfall. Erosion on hillsides and riverbanks can become prevalent where water or wind eat away at bare, unprotected earth that is void of plant life and their root systems. Without a network of roots to hold ground within their grip, soil is at risk of being carried away. Most likely weeds will begin to appear and thrive in all of these difficult growing conditions.

Fortunately, weeds are an excellent reporting tool when it comes to soil health and fertility. Weeds are Mother Nature's way of addressing problems within the soil. Red or white clover loves to establish itself in soils lacking in nitrogen, for example. The clover itself is a nitrogen fixer and, with the assistance of the mycorrhizae and microbes surrounding its roots, clover pulls nitrogen from the atmosphere and leaches it into the ground as its roots perish.

Similarly, dandelions seek compacted swaths of earth for their home. Their long taproots snake their way through hardened layers of topsoil, loosening compacted ground. See? When left to her own devices, nature will try to correct challenging soil conditions. Often it is through human farming activity that weeds are removed, their indications ignored, and the soil ailment goes undiagnosed and unremedied.

Hard clay is a common soil problem. The addition of compost and the proper cover crops can help loosen and restore soil quality.

READ THE WEEDS	
Dandelions	Implies compacted soil, an excess of potassium, or calcium deficiency.
Crabgrass	Prefers poor, dry soil conditions which lack organic matter and fertility.
Wild Violet	Thrives in shady, wet soils. Indicates poor drainage.
White and/or Red Clover	Clover indicates high levels of chlorine, magnesium, and sodium within the soil. Also prefers to grow in areas with low nitrogen quantities as it prefers to fix its own nitrogen in the soil.
Onion Grass	Loves acidic soil with little organic matter.
Purslane	Indicates rich soil with high amounts of phosphorous.
Creeping Charlie	May indicate a lack of organic matter, soil bacteria, and/or nitrogen. High levels of calcium or iron may also be present.
Plantain	Can be found where other plant growth may struggle. Usually a sign of compacted, tight soil conditions.
Bluegrass	Thrive in wet soils with poor drainage.
Knotweed	Grows well in compacted, acidic, infertile soil.
Chickweed	Can be found in fertile, well-watered soil conditions. May indicate poor drainage, over watering, or compacted soil.
Red Sorrel	Indicates highly acidic soil.
Dock	Prefers acidic and wet conditions where bare soil resides. Seeds can last up to 50 years!
Black Medic (aka Yellow Clover)	May mean a nitrogen deficiency, poor soil fertility and lack of water retention within the soil.
Wild Strawberries	A sign of acidic soil.
Queen Anne's Lace	A sign of poor soil conditions with little moisture.
Yarrow	Prefers poor and sandy soil without moisture.

Yarrow tends to thrive in poor and sandy soil conditions, which lack moisture.

The presence of Queen Anne's Lace can indicate poor soil conditions with little moisture.

THE ADDICTION TO SYNTHETIC FERTILIZERS

For years, the cycle of sowing, growing, fertilizing, and harvesting crops has taken place within the confines of conventional agriculture. For many growers the goal was to feed the hungry crops with fertilizers—not to feed the hungry soil. For example, fields of corn were starving for mass amounts of nitrogen that the soil could not necessarily support. To meet the demand, growers fed nitrogen-centric fertilizers, oftentimes in synthetic form.

In addition, growers thought this made sense when it came to absorbing carbon dioxide from the atmosphere and sequestering it within the soil. If a plant was fed mass quantities of nitrogen, it would surely grow more quickly. Faster growth rates meant not just increased production times for farmers, but also higher quantities of carbon absorption.

In theory, this thought process makes sense. However, as early as the 1920s and 1930s scientists discovered that "too much of a good thing" (in this case, the nitrogen needed to sustain these crops), threw the entire ecosystem within the soil out of balance. In a 2007 article in the *Journal of Environmental Quality* entitled "Synthetic Nitrogen Fertilizers Deplete Soil Nitrogen," researchers reference two prewar academic papers noting that the use of synthetic nitrogen fertilizers promoted the loss of carbon and nitrogen from within the soil. The researchers' argument was that nitrogen fertilizer stimulates soil microbes that feast on organic material. Over time, would the starving microbial overconsume crop residues and organic matter? The answer, they found, was yes.

A Holistic Approach

Synthetic fertilizers are used to keep up with the increasing nutritional demand of crop species derived for high produce output. They also are used to replace what many farms have begun segregating over the last century: animals and their manure. Animal manure is an amazing source of organic matter and nitrogen for soils that need replenishment. Once upon a time, most crop fields were spread with fresh or composted animal manure. Farmers simply would capitalize on a free source of readily available fertilizer, the output from their livestock, to fortify their soils. Now livestock often is raised dependent on manufactured animal feed in factory farm situations and feedlots. Many of the very farms that now grow animal feed crops are using synthetic nitrogen fertilizers, breaking the animal-soil nutrient cycle.

These last few pages paint a fairly grim picture of traditional farming practices and their effects on our soils and, ultimately, on our environment. The good news is that anyone can work with Mother Nature to rebuild our soil systems without relying heavily on synthetics.

Applying too much manufactured nitrogen fertilizer causes overactive soil microbes that quickly deplete organic matter within the soil to meet their increasing appetites. Less organic matter directly correlates to less carbon sequestration ability and less organic nitrogen retention. As a result, less organic nitrogen is retained within the soil, and it leaches away. This leached nitrogen ends up in groundwater in the form of nitrates and enters the atmosphere as nitrous oxide (N_2O). This greenhouse gas has roughly 300 times the heat-trapping power of carbon dioxide.

After the soil's organic matter is consumed by microbes and organic nitrogen leaches away, the result is compacted, infertile soil incapable of water absorption and retention. This means an increase in crop irrigation. All the while, growers see a need for more nitrogen as the soil no longer has the ability to retain nutrients. And so a cycle begins.

WHY HEALTHY SOIL MATTERS

Many factors and practices contribute to unhealthy soil ecosystems—but why bother trying to rectify and rehabilitate it? Before I explore a few topics in greater detail, here are a few quick reasons why soil is in the foundation for productive and sustainable agriculture:

- Healthy, nutritionally dense soil can produce healthy, nutritionally dense food for humans and animals.
- Farming for soil health creates a land stewardship relationship between land and grower.
- It fosters carbon absorption and reduces land erosion.
- Sound soil structure is capable of maximum water absorption and improved nutrient cycling.

Thanks to the above, it means you will have better overall land resiliency.

Carbon

As plants pull in energy from the sun, their leaves and roots absorb carbon. Simultaneously, sugars are generated as the plant's food, and oxygen is transpired into the atmosphere. This is the process of photosynthesis. While all plants photosynthesize and absorb carbon, some are more successful at pulling carbon dioxide from the atmosphere than others. Annual crops with a shorter life-span and less woody tissues retain less carbon than their permanent perennial counterparts. (I'll touch on this further in perennial plantings, page 52.) Additionally, the deeper the roots of the plant, the more carbon the plant can withdraw from the air and leach into the lower soil layers.

When carbon is pulled and retained within the soil, this is called a *carbon sink*. A carbon sink is a place where more carbon is stored than is released. Vigorous soil ecosystems that are not disturbed via till methods contain different root depths from varying species and permanent plantings. As a result, the soil layers are replenished with organic matter, are able to absorb massive amounts of carbon dioxide from our atmosphere and assist in reducing harmful greenhouse gases.

Raised bed soil, in addition to in-ground soil, is a mix of organic material and, when healthy, should resemble chocolate cake.

SOIL LAYERS

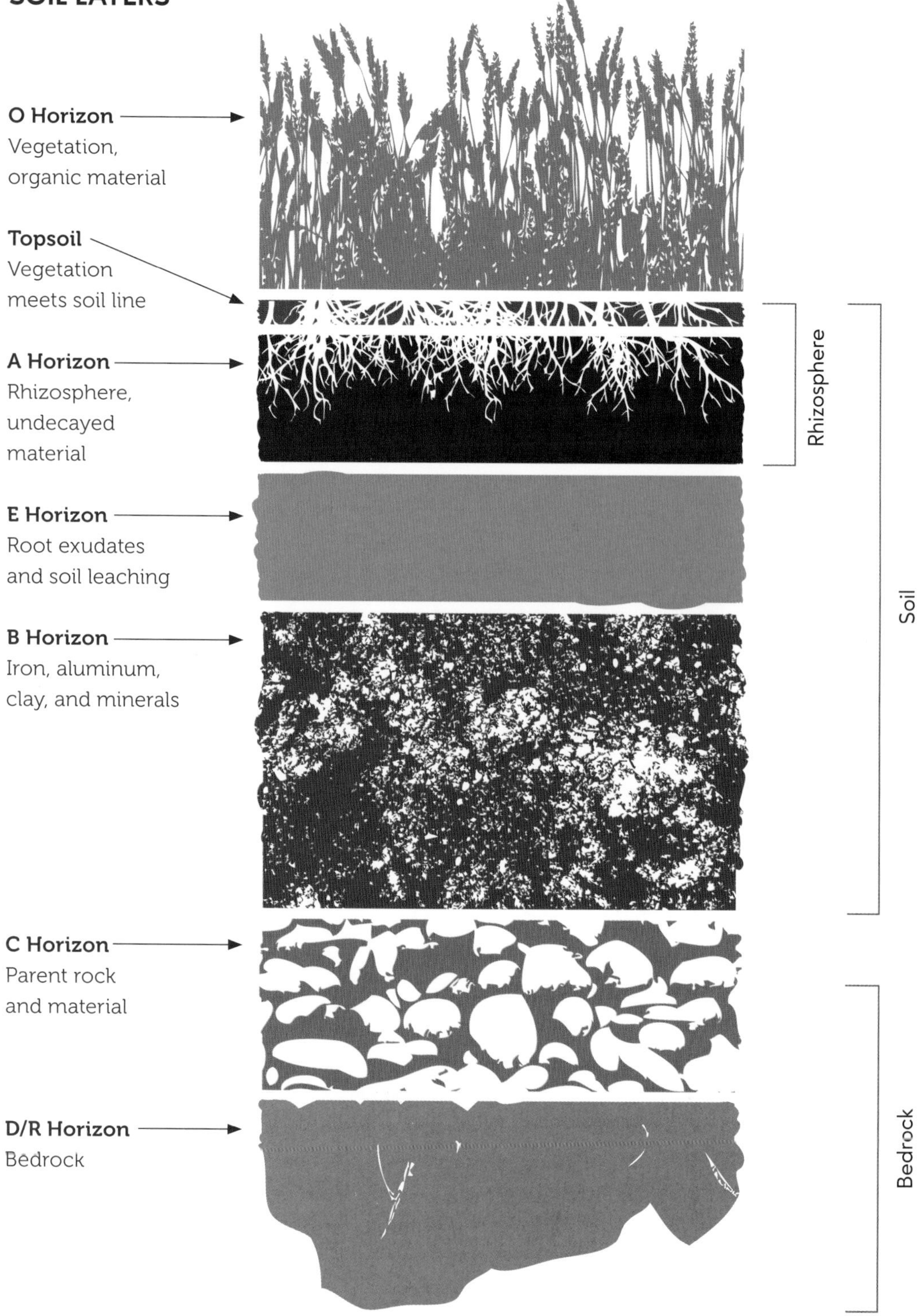

O Horizon
Vegetation, organic material
Topsoil
Vegetation meets soil line
A Horizon
Rhizosphere, undecayed material
E Horizon
Root exudates and soil leaching
B Horizon
Iron, aluminum, clay, and minerals
C Horizon
Parent rock and material
D/R Horizon
Bedrock
Rhizosphere
Soil
Bedrock

Erosion Reduction

Soil top and sublayers are broken down into several sections. The first and top thin layer of soil is called the O Horizon. This layer consists of vegetation growing above the soil line and undecayed organic material. The vegetation here makes contact with the soil at a level called *topsoil*. Directly below this, where plant roots reside, is the section referred to as A Horizon. This layer of earth contains the rhizosphere where microbe, mycorrhizae, and root interactions take place. This also is where humus is built by way of compost and decayed material. Underneath A Horizon is the subsoil layer called E Horizon where root exudates and soil leaching take place. Beneath E Horizon resides B Horizon, where minerals, such as iron and aluminum, and clay are deposited from soil layers closer to the earth's surface. Horizons C, D, and R are the deep layers within the ground where parent material, rock, and bedrock can be found.

If soil is exposed and left bare, its topsoil layers (and their nutrients) are baked by the sun. Hardened earth is conducive to washing away a bit at a time with intense water runoff or carried away as dust in the wind, taking all essential soil nutrients with it. O Horizon, the rhizosphere, and A Horizon are slowly eaten away over time, revealing lower layers. But if vegetative growth, be it crops or natural plantings, is grown within the top layers of soil, the leaves and stems act as a shield, protecting the soil from sun exposure, wind, and rain runoff. The roots, and root clusters from previous plantings left in place, create a network of webs, tightly holding soil and material, keeping it securely anchored.

Water Absorption

Soil is comprised of millions of particles measuring different sizes. Every soil particle falls into one of three categories based on its size: clay, silt, or sand. Clay particles are smallest, measuring under 0.002 millimeters. Silt particles measure

Humus, created from a proper balance of sand, clay, and silt particles, is visible here, creating ideal planting soil.

Despite major flooding in this pasture at least once yearly, forage and vegetation are plentiful thanks to regenerative agriculture. Here is an example of the animal-soil nutrient cycle in practice.

0.002 to 0.05 millimeters. Sand particles are largest, measuring 0.05 up to 2 millimeters. (Anything larger than 2 to 75 millimeters in size is referred to as gravel, which is not considered a soil type.) When you have a balance of all three soil particle types, you have achieved loam.

Well-balanced, porous loamy soil comes from organic matter by way of decaying leaves, varied animal manures, compost, dried grass clippings, and/or shredded tree bark. Loamy soil facilitates the improved infiltration of water. Air pockets created by the three different soil particles coming into contact with plant root systems provide a place for water to go within the soil's layers. As the loamy soil grows and becomes deeper, subsoil layers including the rhizosphere, A Horizon and E Horizon are able to hold and retain moisture in addition to the topsoil. When rainfall is scarce, plants are able to access water stored deeper within the soil's air pockets preventing stress and overall crop loss. And if rainfall is excessive, the soil is able to absorb it like a sponge and hold it for future use.

Nutrient Cycling
You may be familiar with the cycle of water as it falls from clouds in the form of rain, enters waterways, evaporates into the air, and then is held in clouds where it is eventually returned to the earth. Much in the same way that water cycles, so do nutrients within the soil. In an ideal system, carbon is absorbed by plants and leached into the soil via its root system. Nitrogen also enters the soil layers from deteriorating plant material and animal manure, and phosphorous is given to the soil by way of weathering rock. Plants uptake all of these nutrients and provide fodder for animals. Animals consume the vegetation, and their manure is returned to the earth. Plant residue from those consumed crops decays and returns nutrients to the soil. The plant uptakes these nutrients, and the cycle repeats.

Land Resiliency
Carbon and water absorption and retention, nutrient cycling, and erosion reduction create a stable landscape that can be resilient to many forms of natural disasters. Soils fertile with nutrients and hydration are less prone to burning, flooding, and drought. When these dramatic events are present, the land is able to recover and to continue to support vegetation and ecological functions.

My horses graze a pasture home to native grasses and trees. This area floods heavily and is submerged beneath several feet of rushing water multiple times per year. Despite these natural challenges, the land is rich in porous soil structure, which is able to readily absorb water and retain moisture during times of drought. This is all thanks to a wide variety of forageable crop species, animal droppings, pasture rotation, and perennial plant deposits.

THE SOIL NUTRIENT CYCLE

Key // **N:** Nitrogen / **P:** Phosphorous / **C:** Carbon / **O:** Oxygen

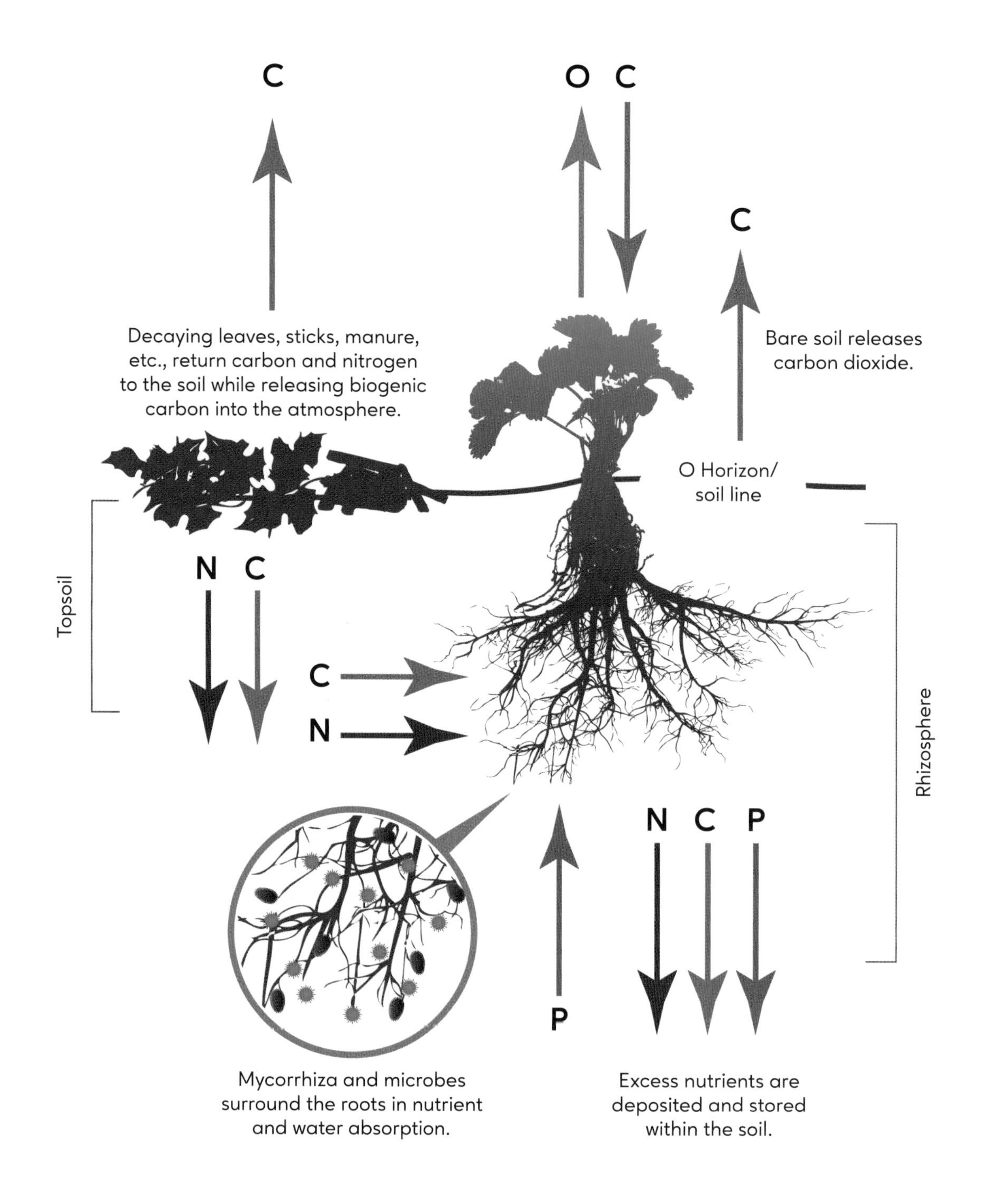

Mycorrhiza and microbes surround the roots in nutrient and water absorption.

Excess nutrients are deposited and stored within the soil.

OTHER NUTRIENT SOURCES

Key // **N:** Nitrogen / **P:** Phosphorous / **C:** Carbon / **O:** Oxygen

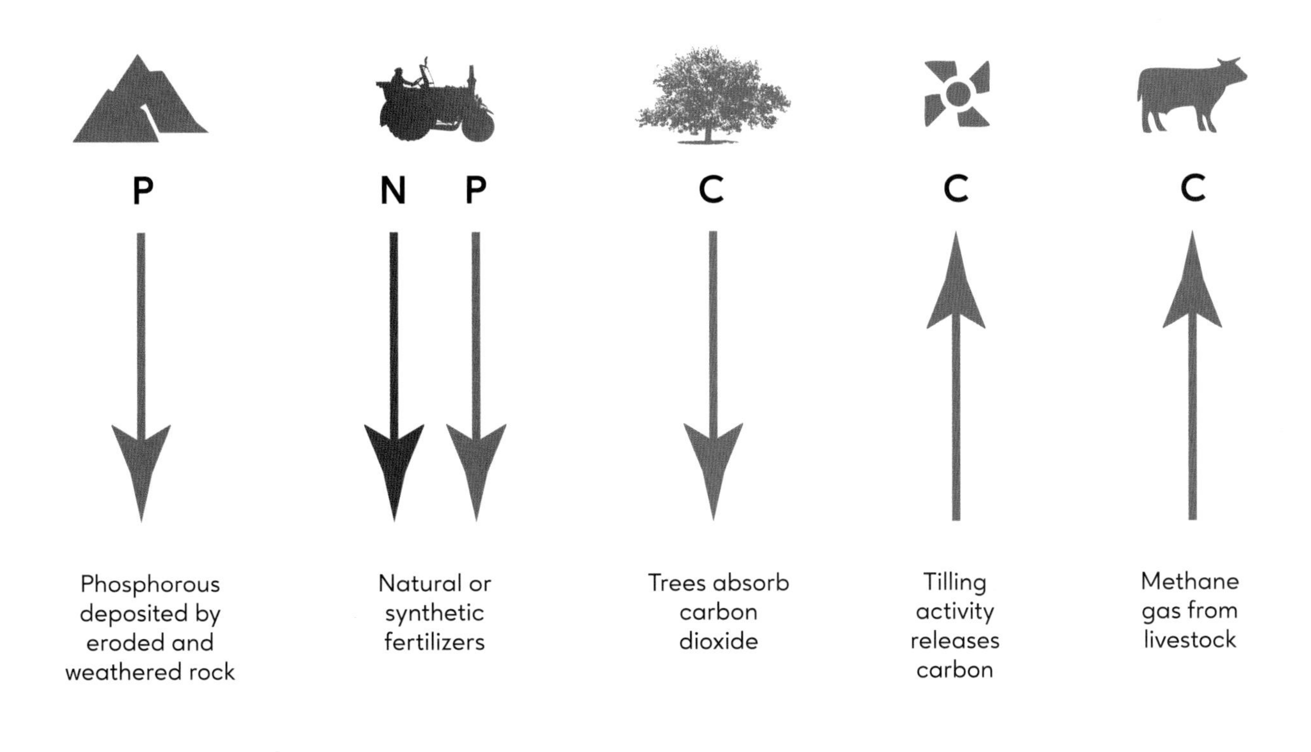

MEMBERS OF THE RHIZOSPHERE

The rhizosphere is the area surrounding the roots of plants where beneficial microbes exist. These microbes help the plant absorb nutrients from the soil, reducing the need for synthetic fertilizers. A teaspoon full of soil contains more microbes than there are people on earth. These integral contributors include the following members:

Bacteria

Fungi

Earthworms

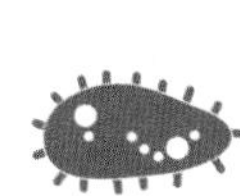
Protozoa

Nematodes

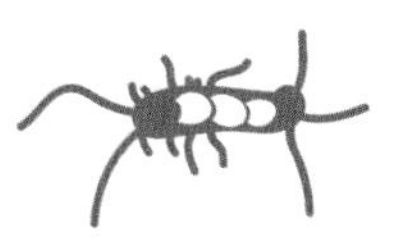
Arthropods

HOW TO REGENERATE AND REPAIR SOIL WITH NATURAL METHODS

Have you ever seen a weed growing through the cracks of a sidewalk? Or an abandoned building surrounding by weeds, vines, and growth? This is nature's way of taking back human-modified spaces. When left to her own devices, Mother Nature will begin to install vegetation and correct manufactured settings in her resiliency. You can help nature to rectify problematic soil conditions by turning to holistic and sustainable farming practices.

Cover Cropping

Many growers are familiar with the concept of cover crops as a means of fixing nitrogen and other nutrients into the soil. If you are unfamiliar, this is the practice where specific plant species are chosen by the farmer or gardener for their ability to absorb nutrients from the atmosphere, loosen compacted soil, and/or uptake nutrients from lower soil layers. Oftentimes, these cover crops are sown, grown, and mowed down just before they go to seed, then they are left to decompose as a form of green, or plant-based, manure. Nutrients the plant absorbed during its life cycle are returned to the earth, and new organic growing material is added by way of decomposed plant matter. If cover crops are grown during the winter, the land is protected from harsh weather, which ensures nutrients stay within the soil until the next growing season. Erosion is prevented as the roots of growing plants hold the soil and water is continuously absorbed.

Anyone can increase the soil fertility, porosity, and topsoil layers by growing not just one, but an entire portfolio of cover crops at once. A myriad of cover crops will deliver myriad nutrients just as nature intended. Cover crops can be grown in vegetable fields, orchard spaces, and animal grazing pastures. They can be grown alongside cash crops in the warmer months as companions, as winter barriers to arm topsoil, or as part of the pasture seed mix for animals to actively graze. Here are a few example scenarios of effective cover cropping.

Animal Grazing

Here at the farm, I feed two Clydesdales on open pasture. They receive supplemental vitamins and amino acids in pelleted form. My horses are not fed a grain ration; I rely on the forage that my grazing spaces provide to meet the majority of my horses' dietary needs. In addition to grass seed, my pastures are sown in the early spring and fall seasons with turnip and radish to reduce compacted soil. Berseem clover, ryegrass, or winter rye also are added to uptake nitrogen, suppress weeds, add potassium, harbor beneficial insects, and improve topsoil.

After four to six weeks, when the cover crops have had time to establish themselves, the horses have access to specific portions of the pasture by way of a rotational grazing schedule. Any crops that are trampled by their 2,000 pounds (907 kg) of body weight are left to decompose in place, creating a green manure deposit. This planting combination meets the digestible energy and crude protein needs of a mature nonworking or lightly worked horse.

Keep in mind that different animal species ingest different vegetation. They also leave different nutrients via their manure deposits to decompose into the soil. Also, left to their own devices, livestock will graze until no vegetation is left when confined to a specific area. Overgrazed forage is void of the nutrients the animal needs to meet their daily dietary requirements. Heavier species such as cows and horses can quickly trample organic material and topsoil layers if not properly managed; this actually hurts the soil more than it helps it. Poorly managed pastures benefit no one.

Cover crops have been carefully selected and are blended by hand in preparation for spring sowing.

Maintaining animal health is just as critical to a successful regenerative farm operation. A healthy animal will create and release the manure needed to feed the soil. In turn, the soil is able to host nutritional forage that feeds the animals. You can capitalize on this nutritional through rotational grazing—but not just by moving an animal from pasture to pasture. Rather, you can circulate different animal species in a follow-the-leader rotation system

The Clydesdales graze on new pasture that's just sprouting in late winter. The horses are removed later in the day when the shortest grass within the area measures 4" (10 cm) tall. High-traffic areas, including gates, surroundings, favorite spots for a roll or dust bath; outbuilding entrances; walkways; and foot paths, can be the most difficult to seed successfully.

through forage spaces. Different species deposit different nutrients and matter onto the topsoil as they feed on varying growth. As the animals come and go, forage is trimmed and has time to rest as it reestablishes. This regrowth period requires photosynthesis and, as a result, carbon is absorbed from the atmosphere. See chapter 4 for more on animal-pasture relationships and rotation example schedules.

Cash Crop Companion Planting

Are you familiar with companion planting? It's one of the foundations of creating working ecosystems when it comes to permaculture. We can group plants together when planning and planting our growing spaces to help crops deter pests/disease, give and take a variety of nutrients to and from the soil, and attract pollinators. Companion planting is also a way to help maximize available real estate. Thinking in terms of growing upward and not outward allows growers to cultivate and harvest more food per acre than if not companion planting at all. An acre of one single crop results in a harvest from one single crop. An acre comprised of various fruits, vegetables, grasses, nuts, and legumes that all support one another results in a much more abundant harvest from that one acre of land. The outcome is a higher crop yield, less human intervention by way of fertilizing and pest control, healthier soil structure, and maximized garden space.

There are a few well-known methods of companion planting for cash crops. The Three Sisters planting method, founded by Indigenous Peoples of the United States, is perhaps the best-known method. It nurtures both plant and soil health, while delivering high yields. With the Three Sisters approach, a corn kernel is planted directly into the soil. After germination, when the corn seedling reaches roughly 6 inches (15 cm) in height, bean seeds are sown surrounding the young corn. Throughout the growing season, the corn acts as a strong support trellis for the beans. In exchange, the beans offer nitrogen to the soil, which the corn plant requires in abundance. Squash or gourd seeds also are sown alongside the thriving corn and bean plants. Their tumbling vines happily roam the soil beneath the corn and beans, keeping the soil cool and moist. This planting concept can work well in smaller growing spaces. Larger fields of corn could be interplanted with other secondary swaths of vegetation. White or red clover is an excellent nitrogen fixer, a nutrient the corn heavily feeds on. Small clover blossoms attract pollinators, harbor beneficial insects, act as a living mulch or groundcover, and provide weed suppression.

Winter Cover

Hairy vetch, field peas, lentils, and cow peas are all nitrogen-fixing cover crops. When sown in the late summer or early fall months, these plants have time to establish themselves before harsh winter weather arrives. These legume crops (when grown alongside non-legumes, such as barley, buckwheat, or rye) work together to improve subsoil and topsoil layers, suppress weeds, replenish nitrogen and phosphorous, and draw carbon into the soil. Come early spring the plot is mowed, and the plant matter is left to decompose in place, creating fresh biomass. New spring seeds are sown directly into the cover, protected from harsh sun and spring rains. This helps to prevent seed washout and assists the new seedlings in staying moist as they begin to grow.

Cover crops are an extremely powerful tool when the farmer selects specific species to address their specific soil challenges. A cover crop combination that works at my farm may not work in another farmer's climate or soil. Use the table on page 49 to create your own custom cover crop recipe to address your soil's challenges, animal grazing needs, and crop nutritional requirements.

COVER CROPS

Crop Species	Hardiness	When to Plant	Contribution
Hairy Vetch	-10°F (-23°C)	early fall	nitrogen, beneficial insects, phosphorous, topsoil health, weed suppression
Crimson Clover	0° to 10°F (-18° to -9°C)	late summer	nitrogen, beneficial insects, phosphorous, weed suppression
Sweet Clovers	-10°F (-23°C)	April through August	subsoil health, phosphorus, nitrogen, weed suppression, topsoil health
Lana Vetch	10° to 15°F (-9° to -15°C)	early spring or late summer	nitrogen, beneficial insects, phosphorus, topsoil health, weed suppression
Spring Field Peas	10° to 20°F (-9° to -6°C)	early spring	nitrogen, beneficial insects, phosphorus, topsoil health
Berseem Clover	20°F (-6°C)	mid- to late summer	nitrogen, beneficial insects, phosphorus, topsoil health, weed suppression
Bell/Fava Beans	20°F (-6°C)	early spring or late summer	nitrogen, beneficial insects, phosphorus, subsoil health
Cowpeas	30° to 100°F (-1° to 38°C)	early summer	nitrogen, beneficial insects, phosphorus, topsoil health, weed suppression
Soybean	40–100°F (4° to 38°C)	spring	nitrogen, beneficial insects, topsoil health
Winter Rye	-40F° (-40°C)	fall	weed suppression, nitrogen, topsoil health, beneficial insects, potassium
Winter Wheat	-25°F (-31°C)	fall	weed suppression, nitrogen, potassium, topsoil health
Triticale	-10°F (-23°C)	fall	weed suppression, nitrogen, topsoil health
Barley	10° to 15°F (-9° to -15°C)	early spring or late summer	weed suppression, nitrogen, beneficial insects, topsoil health
Spring Oats	15° to 20°F (-15° to -6°C)	early spring or late summer	weed suppression, nitrogen, topsoil health, beneficial insects
Radish	20°F (-6°C)	late summer	topsoil health, subsoil health, weed suppression, beneficial insects
Buckwheat	35°F (2°C)	May through August	beneficial insects, weed suppression, phosphorus, topsoil health
Sorghum-Sudan Grass	35°F (2°C)	early summer	weed suppression, subsoil health, nitrogen

Companion plants work together to give and take a variety of nutrients to and from the soil. They also assist one another in repelling unwanted insects, attracting beneficial pollinators, and more.

Decaying organic matter and plant material will be visible on top of the soil. After planting, the soil will be mulched to retain moisture and nutrition.

AMENDING WITH COMPOST

When animals and crops coexist in shared spaces, soil is replenished naturally. Nutrients are returned to the O Horizon, topsoil, and A Horizon by way of animal manure. Added biological material decomposes and new biomass or humus is created. In situations where animals and crops cannot interact, you can bring in fresh or composted manure to facilitate and mimic this natural process.

Horse and cattle manure is loaded with undigested forage seed. If directly incorporated into garden beds and crop fields without the seeds having been burned off by way of hot composting (see page 130), the seeds will begin the process of germination. While this is beneficial for pastures where forage is readily grown, these forage seedlings can crowd out cash crops and compete for much-needed nutrition. For this reason, I use composted manure in my own growing beds, orchard, and crop fields. Fresh manure is spread only in grazing spaces.

You can amend soil without tilling, digging, or disrupting existing soil biomes. Adding a layer of compost no thicker than 2 inches (5 cm) directly on top of growing spaces can dramatically improve soil health. Rain, earthworms, and beneficial bacteria will do the heavy lifting of drawing the freshly laid matter and its nutrients downward. Soil is best topdressed with fresh compost in the spring before new plantings and again in late summer or early autumn before cold-hardy crops are installed.

Loose, moist soil with lots of humus has produced large, quality root crops.

No-Till Method

Tilling is the practice of stirring or overturning soil as a means of preparing an agricultural site for cultivation. Human, horse, or machine-powered tilling all strive to achieve the same goal; loosening soil before planting. Many conventional farmers believed that a loose soil structure meant increased water retention and easier establishment for plant roots. But the opposite is actually true.

Cracks and crevices that have been created in the soil from tilling invite water downward, but the water is traveling at a much faster rate than the soil sublayers can absorb. Air pockets created by root systems and beneficial soil organisms have now been overturned and no longer exist. Water has little place to go and the result is saturated soil with poor drainage.

Just like when unprotected topsoil is left exposed, the solarization, erosion, washout, and nutrient loss of overturned soil is inevitable. The same risks apply to soil layers that are unearthed but with additional effects. Disrupted soil biomes leach carbon dioxide, nitrogen, and other nutrients that they had once absorbed. Roots that are loosened are no longer able to hold soil and their microbes and mycorrhizae interactions are broken. If crops are sown on top of soil layers and mulched with compost, sown into cover crop plantings that have been chopped and dropped, or fixed using a no-till drill, soil structures remain intact along with their nutrient storage tanks, their microorganisms, and their ecological functions.

Perennial Plantings

Perennial plant growth plays a major role in regenerative agriculture. A crop that is sown once and comes back every year with a harvestable yield can do so much for the environment—in addition to providing the farmer with less work and money spent on crop seeds and planting. Perennials are comprised of more fibrous, woody stems and strong plant tissue. Their permanent root structures are vast and run deeper below the soil's surface; this helps improve soil structure. These expansive roots loosen compacted soil and aid in the prevention of erosion, and some perennials mine different nutrients from deep under the soil's surface to help feed their surrounding companions.

The combination of woodier growth and more developed root systems means more carbon is absorbed through photosynthesis and leached deep into the soil. While soil health is important, perennial trees, shrubs, and vegetation provide shelter for wildlife within the ecosystem. They host birds and beneficial insects who help to control pests within the garden and other growing spaces. Rather than focusing on monocrops, try incorporating fruiting trees and caneberries (for example). See chapter 3 for more information on growing and combining perennial plantings.

Flowering bulbs such as daffodils and tulips are perennial choices that benefit pollinators early in the season.

CHAPTER *three* / GROWING

It does not matter if the seed is sown in a permaculture food forest, raised bed, hügelkultur system, container, or directly in the ground: once a seed is planted it is unwavering in its basic requirements in order to grow and thrive. It needs soil, sunlight, and water.

In the previous chapter, I explained that soil health is the basis for a thriving homestead ecosystem. When the soil is vigorous, our animals and crops are as well. In the same way that growers can manipulate their farming practices to facilitate and regenerate soil health, growers can also learn to manage water and utilize optimal sun exposure to encourage plant growth and maintain ecological health.

Crop selection also is hugely important. Learning the crop species that are best suited for your growing climate directly correlates to finished crop size, yield amounts, and overall plant vigor. Creating neighborhoods of companion plants that work together to support one another results in flourishing crops that also give and take a wide variety of nutrients to and from the soil.

« Rather than having synthetic fertilizers applied at planting time, the soil has been conditioned and amended over time. This creates nutrient-dense soil that supports the growth of seedlings.

GETTING STARTED

The first step to becoming a more holistic grower is to identify your growing zone. In the United States, the USDA Plant Zone Hardiness Map is a free resource available online. This map communicates your geographical location's plant hardiness zone relative to the first and last dates of frost. With this information in mind, farmers and gardeners can identify the number of growing days available within a season from spring to autumn.

Plants that require a long growing season will not be suitable for cold climates. Comparatively, cold-loving vegetation would wilt and dry in hotter regions. Learning to grow crops that are suitable to your climate, or even native for that matter, will require less money, energy, and time overall. Crop species already adept to your zone are more likely to thrive without constant intervention. They likely will not require much in the way of water, shade, or extra sunlight beyond what Mother Nature traditionally provides in your location.

Once the growing zone is established, it's time to create growing spaces. Many folks are familiar with in-ground, raised bed, and container gardening.

Containers

Containers are a convenient and portable solution for growing crops or herbs where real estate isn't readily available. Growers can easily fill a planter with potting soil and add any desired amendments to suit the needs of the plants growing inside. The downside to container gardening is that plants will require lots of water. A small space dries out more quickly and, as with terra-cotta or clay, can lose moisture by evaporation through the walls of the pot. Metal growing vessels can become quite warm, heating the soil within. This can damage heat-sensitive roots during the peak of the summer season. Despite the need for more water, pots should always have at least one drainage hole to allow water a means to leave the container. Many crops do not like consistently wet roots and will succumb to rotting and mold if water cannot escape.

Raised Beds

Raised bed gardening can solve many of the challenges that are present when growing crops in containers—if space allows. Raised beds are typically frames constructed of untreated lumber, logs, cinder blocks, stone, or metal. Within the frame, raised bed soil and compost are mixed to provide a growing medium for crops. Ideally, raised beds should measure a minimum of 12 inches (30 cm) high to allow plenty of space for root crops to grow downward. Overall width and length varies, but 4 feet by 8 feet (1.2 × 2.4 m) is a common size as it allows the grower ease of reach into all portions of the raised bed.

Raised beds make things easy. They allow for water to drain away, preventing pooling and soaked plant roots. Their sizes allow for amendments to be added to the soil, and they are an excellent choice when in-ground gardening is not an option. Geographical locations with clay, sand, rock, or wet and soggy terrains are very difficult growing conditions; a raised bed can create an immediate growing space.

Filling a raised bed with pure raised bed soil can be costly, especially for large gardens. My favorite method for filling a raised bed uses biodegradable items from around the homestead at little to no cost. Organic items are used to contribute to the soil's structure as they decay, feeding the plants and the soil as they decompose, just as in nature. Start by placing the constructed raised bed frame on flat ground, in your desired growing location. There is no need to weed or remove the grass beforehand. Next gather boxes and cardboard. Though soy- and vegetable-based inks are commonly used in printing today, boxes with minimal ink coverage are more desirable as less ink will then be introduced to the soil. Be sure to avoid any cardboard with varnishes, foils, or metallic finishes as these can be toxic. Remove stickers, tape, and adhesives because these will not break down readily.

Lay the cardboard flat directly on top of the grass or weeds. The vegetation will be smothered with the contents of the raised bed and, as the vegetation dies, it will return its absorbed nutrients to the soil. Next collect logs and large branches. Mature logs with moisture are further along in the decaying process and will break down within the soil more quickly. Place the logs directly on top of the cardboard, preferably in a single layer to create a somewhat flat surface. This wood will be topped with the next layer, consisting of smaller branches and twigs. Collect decaying leaves, straw, and plant clippings for the following layer and spread evenly. Finally, fill the remaining bed height with compost and raised bed soil. Seedlings can now be transplanted into the bed or seeds may be directly sown. Be sure to cover the top layer of soil with mulch to ensure water retention, prevent soil nutrient solarization, and create a weed barrier. This layered approach to filling a raised bed creates a long-term feeding system: new plants can source nutrients from it, and it ultimately becomes new organic growing material.

In-Ground Gardens

Because raised beds are constructed by filling a frame, this can be limiting for those looking to grow large swaths of crops at a given time. For others, building and filling raised beds may not be an economical solution. In-ground gardening allows the grower to grow without borders. The only limit to the size of the growing plot is determined by the terrain itself. A plot of land with plenty of sunshine, good drainage, and porous, fertile soil will make a productive in-ground garden location. Compost and other desired soil amendments can be added to the soil's surface to enhance soil structure and maintain fertility throughout the growing season. Fencing should be installed to deter animals and wildlife from using the garden as a traffic or feeding area.

To create an in-ground growing space from an area already dense with grass or weeds, the same approach can be taken as when starting a raised bed. There is no need to till, rip out, or remove currently existing growth. Cardboard that is free of stickers, adhesives, heavy ink coverage, and varnishes can be flattened and laid directly on top. Any preexisting vegetation will be smothered as the cardboard is topped with logs, leaves, rotting straw, wood chips, compost, and garden soil. This layering of organic material will decompose and feed both the original soil below as well as the newly installed plants growing above. As the cardboard buried beneath decomposes, it too will contribute fresh organic matter to the garden space.

The hardships that come with in-ground gardening include erosion, pest control, irrigation, and water retention. Soil that is left bare and exposed between crops results in a loss of nutrients, water evaporation, and soil erosion from wind. Pests can be difficult to contain without any sort of physical border or barrier in an open growing space, and they have the ability to freely and quickly affect an entire growing plot. Fencing a large garden or field can be costly and physically laborious. If the plot is large, access to water will need to be provided for consistent irrigation throughout.

CATCHING AND STORING WATER: IRRIGATING MORE EFFECTIVELY AND EFFICIENTLY

Ideally, crops should receive 1 inch (2.5 cm) of water per week in the form of rain, irrigation, or a combination of both. Upon assessing the site of your homestead, you may have decided your growing spaces should reside where water isn't easily accessible. Or perhaps there are other challenges.

Over the past decade, the planet has continued to increase in temperature. Wildfires consume large swaths of land, and droughts affect the plains. In other regions, heavy rainfall is soaking and flooding landscapes. Extreme weather conditions have provided a cause for concern with regard to water; some folks have too much and others too little.

Thankfully, there are some permaculture approaches to water conservation and garden design that can be employed to remedy many challenges. Growing landscapes can be shaped and managed to facilitate the flow and absorption of water both in heavy rainfall events and light routine irrigation.

Bioswales

Healthy soil structure is porous and has the ability to absorb and retain water. Swales take advantage of this concept, and they are common in many growing climates with hot and dry conditions. Swales are comprised of two sections. First, there are the wide indentations that are carved with gently sloping sides, conforming to the natural shape of the land. The second component is the berm, or mound, placed on the downhill side of the swale. This is where the water ultimately is absorbed.

These wide, sloping cavities slow rainwater and create a space for water to pool. In extreme rainfall events, water that would otherwise leach away can be slowed and stored. As the rainwater flows, it is filtered by the growing crop vegetation that removes sediments and toxic elements. The plant growth also prevents erosion and holds the soil in place. This filtration system and erosion-control method is what make these swales different from their grassy-covered counterparts; these holistically minded versions are called *bioswales*. At the bottom of the bioswale trench, water accumulates, as does organic matter carried by the moving water from the gently sloped sides. The result is a fertile, moist location for water-loving plant life to thrive.

Fruit trees, berry bushes, and caneberries are great candidates for growing in bioswale berms. Because they are perennials and have sophisticated root systems, they absorb a great deal of water and hold soil in place well. Here are a few ideas about where you can incorporate a bioswale on your homestead.

Catchment Ponds

Effective rainfall is the amount of rainfall that the land can readily absorb and hold during a rainstorm. Any excess water runoff is considered ineffective. Rather than let the water escape, a catchment or retention pond can be created. Stormwater can be routed through a series of ditches or pipes from higher elevation to an excavated pond, located at lower elevation. The pond would retain water year-round and serve as a permanent source of hydration for crops during dry seasons. Vegetation surrounding the pond filters out pollutants and any unwanted debris.

Although artificial, a retention pond can serve many purposes for the homestead in addition to water storage. Pond-loving critters, such as frogs,

BIOSWALE INSTALLATION IDEAS FOR THE HOMESTEAD

If you already have a separate growing space for the majority of your crops, a bioswale can be easily incorporated to create a secondary opportunity for plants. These swale ideas capitalize on nutrients that may be leeching away from already existing structures on the homestead.

Key // **N:** Nitrogen / **P:** Phosphorous / **K:** Potassium

CHICKEN COOP BIOSWALE

The manure of chickens, ducks, geese, and other backyard poultry is a naturally balanced fertilizer. High levels of nitrogen, phosphorous, and potassium from the manure can be used to fertilize crops by placing a bioswale adjacent to a coop or chicken run. A border of flowers or herbs is the first level of filtration immediately surrounding the structure. When it rains, water runoff from the coop pools into the swale. Crops and vegetation planted on the berm of the bioswale gain nutrients from the soil.

PASTURE BASE BIOSWALE

Chances are that your pasture space is not perfectly level. Installing a bioswale on the lower slope of the pasture fence can provide nutrient runoff to crops. The manure of horses and other livestock contains high levels of nitrogen. A border of flowers or herbs can be planted as the first level of filtration immediately surrounding the fence. When it rains, water runoff from the pasture pools into the swale. Crops and vegetation planted on the berm of the bioswale gain nutrients from the soil.

NOTE

Due to the pathogens and bacteria in chicken and other uncomposted animal manure, it's best to avoid planting leafy greens and vegetation without a skin that will be removed. Also avoid planting any potentially toxic herbs and flowers surrounding animal structures.

are excellent for fly and mosquito control. When stocked with fish, mosquito larvae are not able to accumulate. Plants that tolerate wet roots, such as marsh mallow, rice, and elderberry, are excellent candidates for circumventing the pond's perimeter.

Rain Barrels

Rain barrels are quite popular in many urban gardens. It's a simple, low-cost solution for collecting rainwater that doesn't take up much real estate. A rain barrel is a large container placed next to a downspout. Stormwater runs from the roof gutter down the spout where it is then deposited into a storage tank. Many rain barrels are outfitted with a spigot pairing for a hose line. This way, water that would otherwise be lost is supplied directly to the garden.

Rainwater is relatively pure. Because it is not treated with chlorine or chemicals, it's perfectly safe for plants. Keeping the barrel in a shady location is best as rain barrel water tends to become contaminated after about one week due to sunlight exposure, animals, and insects.

Ollas

Some growing spaces may receive water inconsistently due to lack of access or rainfall. An olla can help with this challenge. Ollas are clay or terra-cotta containers, much like a pot or jug, planted between crops within the soil. A watering can or hose is used to pour water into the container through an opening at the top which is exposed at the soil line. After filling with water, the opening is covered with a lid to ensure there is no water loss due to evaporation. Because clay and terra-cotta are porous, the water leaches from the container into the immediate surrounding soil slowly over time. When installed among growing crops, the roots of the plants seek the hydrated soil surrounding the olla.

Ollas have been said to offer water to plants as far as 3 feet (1 m) away once crop root systems are fully developed. Because the majority of the container is planted beneath the ground, crops with deep and fibrous root networks are best. Melon, squash, pumpkins, tomatoes, and peppers can benefit from interplanting with an olla. Shallow-rooted vegetation, such as lettuce and spinach, could intake water from the olla neck when planted within just a few inches of the water spout opening.

Complex drip line systems, soaker hoses, and even sprinklers can be costly. Sometimes the landscape may make it altogether impossible to install water lines. This can be challenging, but it doesn't mean that growing crops or forage is impossible. Turning to water storage containers, and even to the landscape itself for water management solutions, can ensure crops are receiving the hydration they require.

Too little water is harmful for plants, as is too much. Most crop vegetation prefers well-draining soil and hates to have its roots constantly submerged in water. To avoid soggy soil, improve structure using the methods mentioned in the previous chapter. Also avoid planting crops in boggy, low-lying locations where standing water is prevalent.

Capitalizing on Sunlight: Permaculture Garden Design

When choosing a site for growing, it is important to remember that not all plants were created to thrive in full sun. In nature the tallest plants, such as trees, receive the majority of sunlight. Items that grow underneath the tree's canopy receive filtered light, and those that grow on the forest floor are shaded. For example, companion planting pairings tell us that lettuce prefers to grow in the shade of taller vegetation, such as tomatoes. If left exposed to the harsh sunlight of summer, the tender leaves of greens can fall victim to scorching. Rather than grow greens in a large swath of landscape and then cover them with shade cloth for protection, a growing space can be designed that takes advantage of sunlight for the plants that truly need it to thrive. Shade-loving items, such as lettuce, can then be interplanted in between to better suit their needs. Ecological garden design can offer not just sunlight optimization, but also water and temperature control to plants.

I have begun installing several food forest-style gardens throughout my property. With guardian dogs present on the homestead, deer and unwanted visitors do not threaten our unfenced crops. Forest landscapes keep the soil cool and better hydrated with their layered growth structure.

MANDALA GARDEN DESIGN

Key // **A:** Pathways / **B:** Focal Point / **C:** Tall Exterior Crops

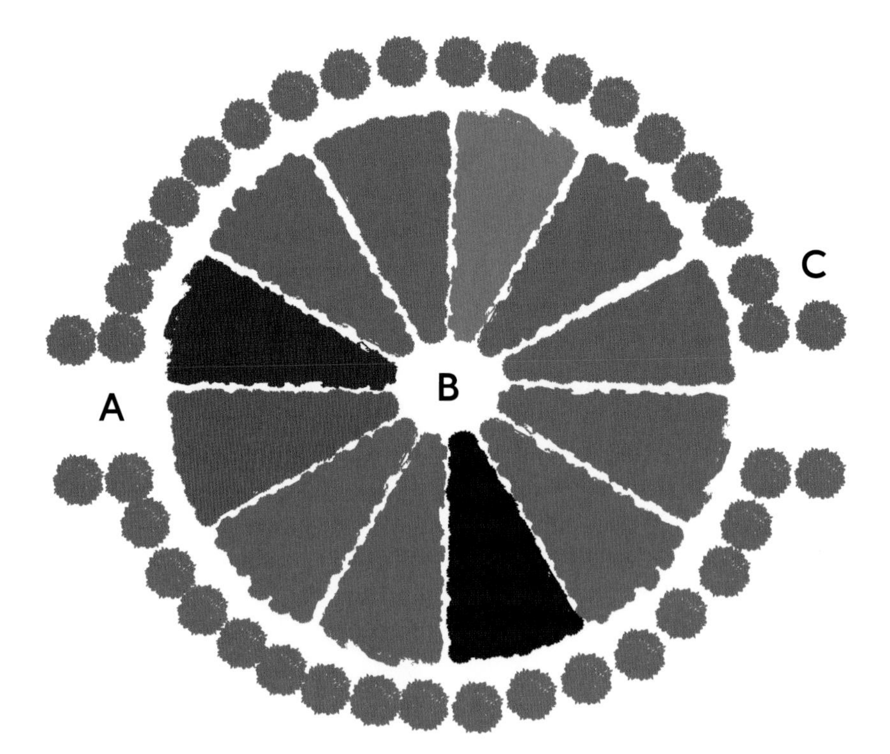

A mandala garden is a circular garden layout in which raised beds or in-ground plantings are arranged in a spiral, floral, or wheel pattern. The garden beds are constructed between narrow pathways, which allows the grower easy access to all plants from all sides. In addition, this arrangement best capitalizes on available garden real estate. When plants are laid out in a circular fashion, with the tallest crops on the outermost perimeter of the garden, the temperature and climate are better insulated within the centermost beds. This creates a microclimate of warmer temperatures inside the mandala.

A small fruiting tree, such as dwarf apple or citrus, can be planted within the center as the focal point of the garden. As the tree grows taller, it will cast shade on certain beds throughout portions of the day. This reduction of sunlight is useful when growing partial-shade-loving crops such as greens and brassicas.

Ideal Crops for Mandala Gardens

A wide variety of crops including corn, sunflowers, pole beans, and vining squash with trellises on the perimeter. Smaller heat-loving crops such as tomatoes and peppers do well within the center.

KEYHOLE GARDEN DESIGN

Key // **A:** Retaining Wall / **B:** Compost Bin / **C:** Planting Area

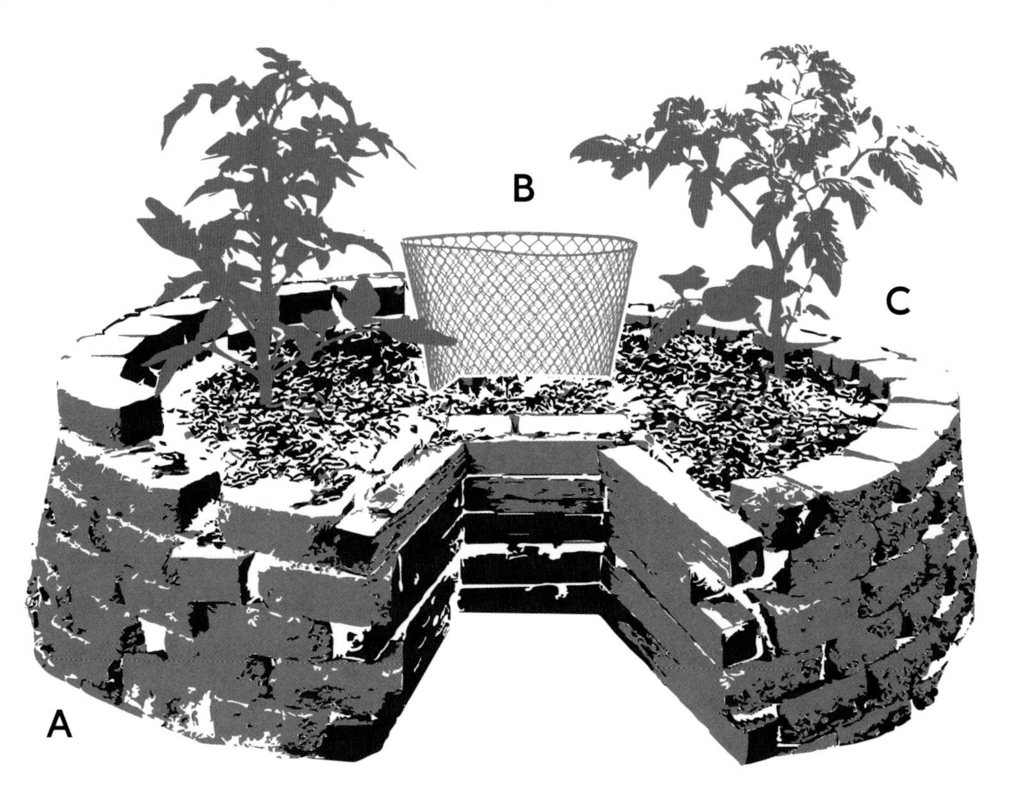

A keyhole garden is a circular raised planting bed fortified around its exterior with a retaining wall of either stone or lumber. A notch is left open when constructing the raised bed to allow the grower access to the center of the garden bed. At the epicenter, a cage or wire basket is created for compost deposits. As the keyhole garden bed is watered, nutrients from the compost within the center leach into the soil. This direct access to decomposing matter results in nutritious soil without having to haul manure or compost to the planting area.

AS THE CONTENTS OF THE COMPOST BIN DECOMPOSE, NUTRIENTS ARE GIVEN TO THE SOIL DURING IRRIGATION.

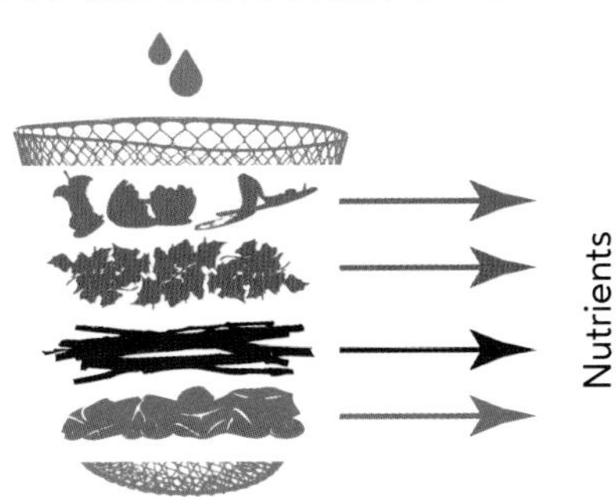

Ideal Crops for Keyhole Gardens

Leafy greens, herbs, onions, carrots, radishes, turnips, tomatoes, peppers, eggplant, garlic, and flowers

SWALE/TERRACE GARDEN

Key // **A:** Berm / **B:** Variety of Foliage / **C:** Excavated Swale

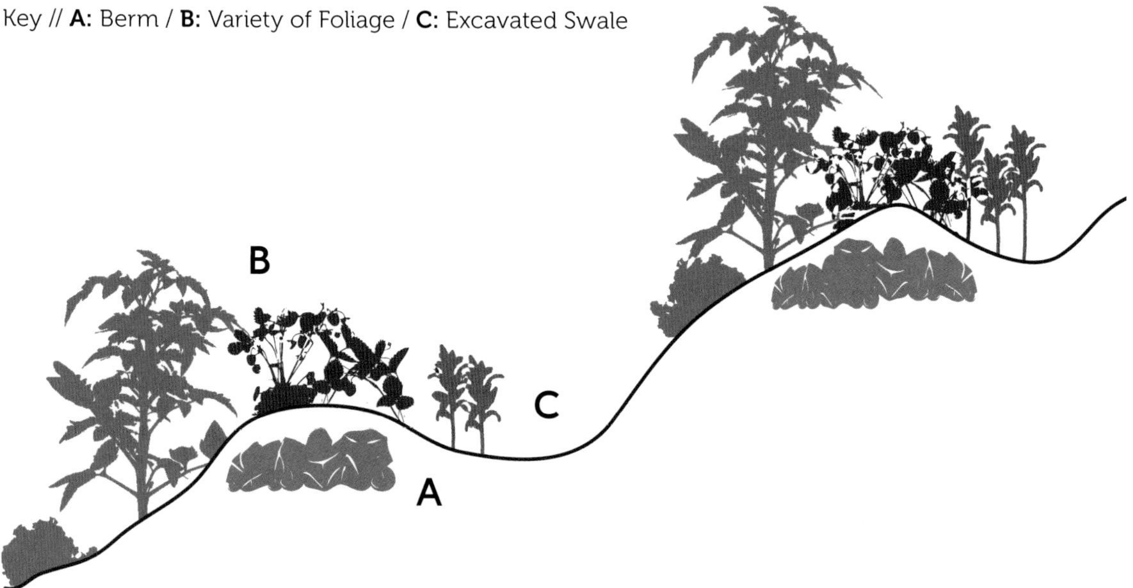

Swales are ditch-like troughs dug from the soil to catch pooling water and slow rain runoff. The removed soil is used to build the adjacent berms in addition to rocks, logs, and other organic material. Water is absorbed into the ground from the swale and hydrates the surrounding crops. Only water-loving vegetation should be grown inside the swale ditches since this area retains the most moisture. Typical vegetable crops like lettuce, tomatoes, and strawberries can be grown on the outer portion of the berm.

HOW A BIOSWALE FUNCTIONS

As rain falls onto the landscape, it is slowed and pools inside the swale ditches. Over time the waters absorbs into the soil and hydrates all the surrounding crops and vegetation.

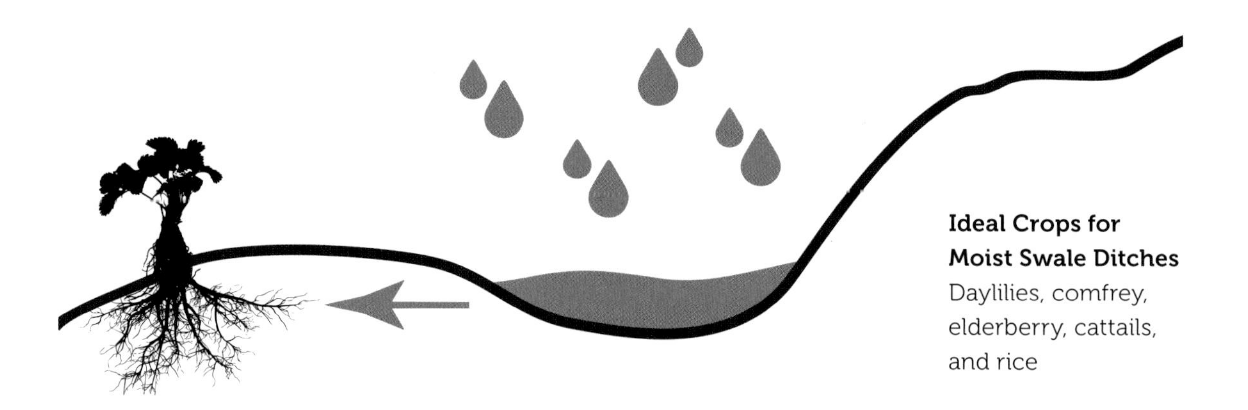

FOOD FOREST GARDEN

Key // **A:** Overstory / **B:** Midstory / **C:** Understory / **D:** Vining Plants / **E:** Groundcover / **F:** Outer Growth / **G:** Companion Plants (Herbs and Flowers)

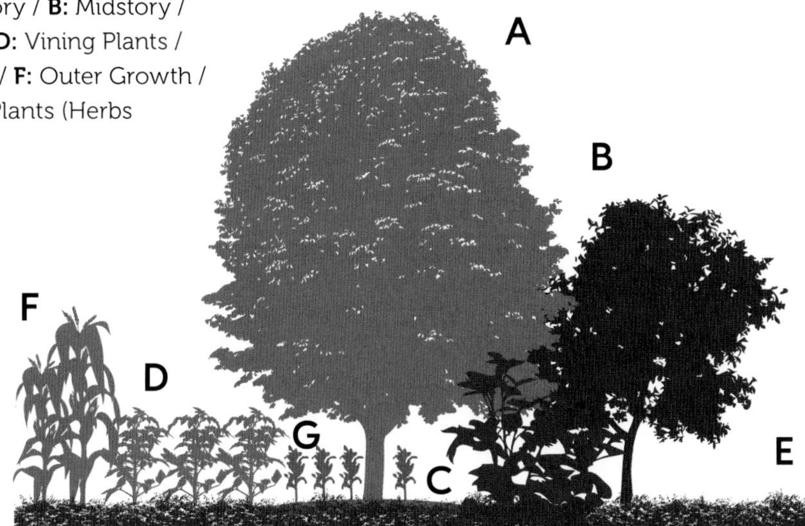

A food forest is an ecosystem approach to planting with various layers of vegetation, just as found in nature. Companion plants come together to attract pollinators, repel pests and disease, absorb nutrients from the air, uptake nutrients and make them more readily available from deep within the soil sublayers, and naturally suppress weeds and mulch the food forest floor. A variety of plants serve a variety of functions while producing a harvest for the grower. A food forest offers more food per acre than any other planting method.

FOOD FOREST MEMBER EXAMPLES

Overstory: Maple, Chestnut, Black Locust, Full-Size Fruit Trees

Midstory: Dwarf Fruit Trees and Nut Trees

Understory: Elderberry, Gooseberry, Raspberry, Currents

Vining Plants: Grapes, Hardy Kiwi, Tomatoes, Honeysuckle

Groundcover: Strawberry, Mint, Wild Violet, Clover

Outer Growth: Full Sun Understory such as Corn

Ideal Companion Plants for Forests

Comfrey, mint, lavender, calendula, marigold, lettuce, daylilies, nasturtium, chives, daffodils, lemon balm, borage, cosmos, rosemary, thyme, oregano, parsley

This garden is inspired by the food forests that have been cultivated in cultures throughout the world for centuries. Here, pumpkins and gourds are growing amongst black walnut trees as well as Russian olive. As the pumpkins and gourds seek sun, they climb the tree branches. These branches create a natural trellis support system.

CHOOSING PLANTS

Most folks recognize that a palm tree is not suited to grow in northernmost climates: it requires ample sunlight, high temperatures, and year-round warmth. When exposed to cold and inadequate sunshine, it would surely suffer and likely die. Many new gardeners don't realize that virtually all crops have their own unique requirements, just as the palm tree does. Though not necessarily tropical in nature, different crop varieties are suited for varying climates, sun exposure, daylight hours, water amounts, and temperatures.

The first fundamental to take into consideration when deciding which plants to grow is seasonality. Some crops are cold-loving, meaning they thrive in temperatures above 32°F (0°C) and below 55°F (13°C). Some crops even prefer temperatures below freezing! Contrary to popular belief, these crops not only love the cold but actually taste better when exposed to their ideal temperatures as they are able to produce more sugars. A common mistake new growers make is trying to grow broccoli, cauliflower, or other members of the brassica family in the height of summer resulting in smaller crop size, bitter flavor, or nonexistent harvests. When exposed to high temperatures, a broccoli or cauliflower plant will quickly go to seed by sending up a flower stalk. This is called "bolting." If a broccoli plant does not have enough time to develop in colder temperatures before reaching the height of summer, it will not develop a proper crown or broccoli head. This is because brassica crops are cold-loving, and they should be grown in spring and fall. Spring and fall gardens can include a wide variety of vegetation in addition to broccoli and cauliflower including turnips, peas, radish, Brussels sprouts, carrots, arugula, spinach, kale, lettuce, celery, and bok choy. Tomatoes, peppers, tomatillos, corn, beans, and pumpkins on the other hand are heat-loving crop varieties. They are best planted after all danger of frost has passed when temperatures are consistently warm or above 65°F (18°C). Some of these crops can be started indoors in seed trays in colder climates while others prefer to be directly sowed. Seed packets are an excellent resource for stating which crops can be started indoor or outdoors and when, seed-sowing depth, and overall growing days to harvest.

Seed starting allows the grower to get a head start on vegetables that require a longer growing season. As soon as temperatures allow, and after the threat of frost has passed, crops started indoors can then be "hardened off" before being transplanted to their permanent growing locations outdoors. Hardening off is the process of acclimating a crop grown indoors to living full-time outdoors. Seedlings can be placed outdoors in a shady location for one hour when temperatures allow, then returned inside. Each day, over the course of ten days, increase the amount of outdoor and direct sunlight exposure by one hour until the plant can comfortably reside outdoors all day. This gradual exposure to sunlight, wind, and rain helps to condition the protected plant for outdoor life.

In addition to temperature, daylight hours can affect some crops when it comes to their growth habits. Onions, for example, are categorized into three groups: long day, intermediate, and short day. Long-day onions are species that require long periods of daylight, typically found in USDA Hardiness Zones 6 and below. Roughly fourteen to sixteen hours of sunlight are required for these onions to create proper bulbs. In regions where there are fewer daylight hours, such as in Zone 7 and higher, short-day onions are appropriate as they prefer just ten to twelve hours of daylight. Intermediate onion varieties (sometimes referred to as day-neutral) can be grown in any region, but thrive specifically in growing Zones 5 and 6, as this where a day length average of twelve to fourteen hours is reached. As mentioned at the beginning of this chapter, you can find your growing zone number by referring to the USDA Plant Hardiness Zone Map found online.

Crop selection can be overwhelming, especially for the new grower. If temperature requirements, water needs, sunlight hours, and daylight length feel like too much to process, try asking yourself, What do I like to eat? Choose one crop variety you often purchase at a grocery store or farmers' market and learn the needs of just that one crop. When you feel like you have a good understanding of the first, add a second and so on. Eventually, you'll have an entire garden.

Selecting Crops for Storage Ability

Many of the crops that I grow are selected and planned around preservation in addition to their permaculture contributions. As someone passionate about reducing my carbon footprint and sourcing food sustainably, I don't want to limit homegrown food consumption to just the warmer growing months. I desire backyard produce year-round. The majority of what I grow has been chosen for its ability to store well as canned, dried, frozen, or whole goods. When provided with proper storage conditions, such as proper packing where applicable, temperature, and humidity, the following crops have proven themselves to be shelf stable.

Crop Varieties for Storage Stability	
Yellow Onions	Ailsa Craig, Bridger, Copra, Cortland, Patterson, Pontiac, Talon, Danvers Yellow Globe, Yellow Sweet Spanish
White Onions	Southport White Globe, Stuttgarter, White Sweet Spanish
Red Onions	Brunswick, Red Bull, Red Creole, Red Wing
Potatoes	Elba, Katahdin, Red Chieftain, Yukon Gold, Burbank Russet, German Butterball, Yukon Gem, Rose Finn Apple Fingerling, Russian Banana Fingerling, Red Pontiac, All Blue, Kennebec
Garlic (softneck best for storage)	Inchelium Red, California Softneck, California Early, Italian Loiacono, Silver White
Sweet Potato	Beauregard
Carrots	Nantes, Chantenay, Imperator, Danvers
Beets	Cylindra, Flat of Egypt, Pacemaker III, Pablo, Boro
Shallots	Chicken Leg (Zebrune), Dutch Yellow
Paste Tomatoes	San Marazano, Amish Paste, Roma, Big Mama, Golden Mama
Cherry Tomatoes (paste and preservation)	Sun Gold, Super Sweet 100, Yellow Pear
Slicing Tomatoes (paste and preservation)	Black Krim, Rutgers, Marglobe, Ace 55
Peaches	Free stone varieties
Apples (whole storage)	Granny Smith, Red Delicious, Golden Delicious, Gala, Empire, Winesap, Fuji
Apples (canning)	McIntosh, Cortland, Fuji, Braeburn, Jonagold, Granny Smith, Golden Delicious, Pink Lady, Jazz, Honeycrisp
Canning Beans (bush varieties)	Black Turtle, Gold Mine, Blue Lake 274, Topcrop, Tendercrop, Contender, Provider, Strike, Improved Tendergreen, Refugee, Stringless Green Pod
Pole Beans	Blue Lake FM-1, Kentucky Wonder
Squash (whole storage)	Spaghetti, Waltham Butternut, Anna Swartz Hubbard, Golden Hubbard, Sibley, Musquee de Provence, Marina di Chioggia, Queensland Blue, Dutch Crookneck, Australian Butter

Consider Adding Companions

In chapter 2, I introduced the concept of companion planting (see page 47). Essentially crops are paired or grouped together to form symbiotic relationships. Orchard guilds are an example of sophisticated companion planting systems (see page 159). Typically, a guild includes the orchard tree surrounded by crops that perform at least six different functions. Helpful vegetation performs weed suppression, fixes nitrogen, and acts as a living mulch in addition to all of the tasks mentioned above.

These concepts can be applied to small spaces—even container gardens—and large areas, such as commercial farms and

Companion Combinations	
Tomatoes plus Lettuce	Lettuce grows well in the shade provided by tomatoes. The lettuce acts as a soil cover, reducing weeds and keeping the soil moist around the tomatoes.
Broccoli/Cauliflower plus Garlic or Sage	Garlic and/or sage deters cabbage moths for the broccoli.
Cucumbers plus Radish	Radishes deter cucumber beetles. In return, the cucumber plant provides speckled shade for the radish tops planted beneath.
Pumpkin/Squash plus Nasturtium	Sow 2 to 3 nasturtium seeds around each seedling after sprouting to deter squash bugs/borers. In return, the nasturtium attracts pollinating insects to increase the productivity of pumpkin/gourd plants.
Roses plus Garlic	Interplant garlic around roses to deter fungal diseases. Some folks say that in return, the garlic encourages a stronger fragrance from the rose blossoms!
Peas plus Nightshade crops	Peas give nitrogen to the soil so are great for heavy-feeding nightshade plants, such as tomatoes or peppers.
Marigolds and Calendula	Marigolds and calendula (pot marigolds) are compatible with most vegetable crops. Use around borders to attract beneficial pollinators and deter rabbits.
Strawberries plus Mint or Borage	Mint acts as a pest deterrent for ants who happily feast on strawberries. Mint flower blossoms are also loved by predatory insects, such as wasps. The tall mint hides the berries from birds. Strawberries also benefit from being interplanted with borage to attract pollinators and enhance berry sweetness.
Broccoli/Cauliflower plus Crimson Clover	Crimson clover acts as a living mulch to attract predatory insects to feed on cabbage moth larvae.
Dill, Fennel plus Melon	The flower blossoms from dill and fennel attract beneficial pollinators that increase the yields of melon plants.
Leafy Greens (Lettuce, Spinach, Kale) plus Alyssum	Sweet alyssum is a great green manure that can be cut and left to decompose in the soil. It also attracts hover flies, which feed on aphids. These aphids often ruin leafy green crops.
Potatoes plus Cabbage	Potato blossoms attract predatory insects that will feed on the cabbage moth larvae infecting cabbage plants. In exchange, cabbage is said to improve the flavor of potatoes when interplanted.

orchards. Remember the Three Sisters example referenced in the previous chapter (page 47)? That is a companion planting system. The approach is simply about moving away from monoculture (planting row upon row of one single species, such as corn). Instead, many crop species are planted together. This practice is called polyculture and results in a wide variety of nutrients being pulled from and deposited into the soil. Plants with varying root depths and of varying sizes improve soil structure at multiple levels that benefit shallow items, such as legumes and leafy greens, to heavily submerged growth from trees and shrubs.

Barn cats are helpful in the garden. They scare away birds looking to steal seedlings from seed trays and freshly sown crops.

Helpful Flowers

Flowers are one of the most underrated tools for the garden. They are useful for so much more than cutting for bouquets and aesthetics. Flowers attract predatory insects, such as wasps, along with pollinators; this increases crop pollination, and thus crop yields. They are useful in improving soil health and repelling insects. When properly companion planted, they can help vegetables better access nutrients. Here are some flower varieties that I've come to favor for their multifaceted functions.

Sunflowers

- Food for honey bees and wild pollinators
- Edible seeds
- Trellis support for beans and small squash
- Deep roots loosen soil and pull nutrients upward
- Provide shade for lettuce when interplanted

Nasturtium

- Edible blossoms
- Deters squash bugs, aphids, beetles, and cabbage loopers for surrounding plants
- Attracts pollinators and predatory insects
- Thrives in poor soil
- Trellis with zucchini, cucumber, pumpkin, and squash

Marigolds (include Calendula/Pot Marigolds)

- Repels whiteflies
- Repels bad nematodes
- Attracts pollinators
- Edible blossoms (French are some of the best tasting)
- Repels rabbits

Sweet Alyssum

- When grown as a cover crop, can be mulched right into the soil for a nitrogen boost
- A living mulch that cuts down on weeds
- Edible leaves and flowers
- Attracts beneficial insects

Borage

- Attracts pollinators
- Replenishes nectar supply in twenty minutes (a super feeder for honey bees)
- Leaves and blossoms are edible
- Adds trace minerals to soil
- Repels hornworms

Companion plants, or guild members, have been carefully chosen for their contribution to this apple tree.

Incorporating Perennials

Imagine planting a seed once and being able to harvest food from that plant for years. Gardening often is thought of in terms of annuals; corn, tomatoes, peppers, etc. These items are planted once, harvested and then they perish. To receive a second harvest, annual crops must be planted and tended to all over again. When thinking about permaculture and holistic homesteading, it's essential to start thinking in terms of perennials; plants that return year after year.

Perennial plants require less work over the course of their lifetime. There is the initial planting phase; but thereafter, only watering (if any) and general maintenance, such as weeding, is required. From that first initial planting, harvests can be reaped for years, often growing larger in abundance than the year before. In addition, the contribution that perennials make to soil structure has already been discussed in chapter 2. Carbon and nitrogen are not just pulled and sequestered within the soil due to perennial root systems; perennials also absorb and withhold carbon in their woodier plant tissues.

Fruit and nut trees are excellent companion plants as they offer filtered sunlight to smaller shrubs and vegetation below. Because permanent plantings become fixtures within growing spaces, they offer insects and wildlife a place to call home. Owls, hawks, and migratory birds create their nests within the branches. They begin to feed on smaller pests, such as mice, voles, moles, squirrels, and pesky insects that often wreak havoc on crops.

Instead of thinking short term, I personally had to retrain my brain to think of growing food as long term. Perennial crops have now taken precedence in my homestead landscape and annuals are secondary. Here is a short list of common perennial crops that come back every single year.

- Apples
- Artichokes
- Asparagus
- Blackberries
- Blueberries
- Cherries
- Chives
- Collard Greens
- Cranberries
- Elderberries
- Grapefruit
- Grapes
- Goji Berries
- Gooseberries
- Horseradish
- Jerusalem Artichokes
- Kale
- Lemons
- Lemon balm
- Lemongrass
- Limes
- Loganberrie
- Lovage
- Mango
- Mint
- Mulberries
- Nectarines
- Olives
- Oranges
- Oregano
- Peaches
- Pears
- Persimmon
- Pomegranate
- Plum
- Quince
- Radicchio
- Ramps (Wild Leeks)
- Raspberries
- Rhubarb
- Sage
- Sorrel
- Strawberries
- Thyme
- Wine Berries

HOW MUCH FOOD TO GROW PER PERSON

Sustainable agriculture is still agriculture: the act of working with the land to produce food. A question I often receive is, How much food should I grow per person? The answer is different for every individual and household. What are your likes and dislikes? What do you eat regularly or purchase at the grocery store on a weekly basis? Therein lies your answer for what to cultivate. For example, I make pizza and pasta weekly and that requires one to two pints (473 to 946 ml) of tomato sauce per meal. Personally, I grow a minimum of thirty-five tomato plants each year because I require enough fruit for fresh eating in the warmer months, and then additional bushels for the preservation of tomato-based canned goods for year-long consumption. The number thirty-five was what I calculated over time, based on my own experience.

If food is grown as a supplement to grocery store shopping, fewer plants per person are required. But if a family is looking to truly become self-sustaining with homegrown goods, many more crops need to be planted. Increase plant quantities even more if your goal is to sell produce at a farm stand or market, if you intend to share with friends or family, or if you'd like to donate to a local food pantry. The following is a general guideline for common crops with regard to how many plants to grow per individual. These numbers are based on crops grown in fertile soil with ideal growing conditions. Feel free to use this as a starting point; increase or decrease these numbers based on your personal preferences.

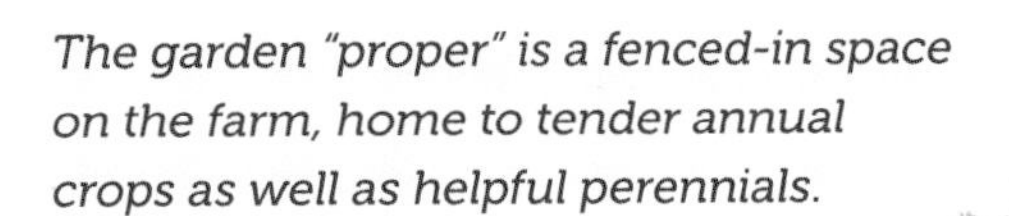

The garden "proper" is a fenced-in space on the farm, home to tender annual crops as well as helpful perennials.

CROP QUANTITIES PER PERSON

Crop	How Many Plants Per Person for Fresh Seasonal Eating	How Many Plants Per Person for a One Year Supply
Apple	1 tree	1 tree
Asparagus	5–10 plants	15–20 plants
Beets	15–30 plants	50 plants
Blackberries	5–10 canes	50 canes
Blueberries	3 bushes	8 bushes
Broccoli	3 plants	15 plants
Brussels sprouts	2 plants	8 plants
Bush beans	12–15 plants	30 plants
Carrots	48 plants	200 plants
Cauliflower	3 plants	15 plants
Corn	10 plants	25 plants
Cucumber	1 vining cucumber or 2 bush varieties	3 vines or 6 bushes per person
Eggplant	2 plants	5 plants
Garlic	10 plants	25 plants
Kale	4 plants	10 plants
Leaf lettuce	24 plants	n/a—cannot be preserved
Melon	1–2 vines	5 vines
Onion	12 bulbs	50 plants
Peach	1 tree	1 tree
Pear	1 tree	1 tree
Peas	15–20 plants	40 plants
Pepper	3–5 plants	9 plants
Potato	10 plants	30 plants
Radish	10–15 plants	75 plants
Raspberries	5–10 canes	50 canes
Strawberries	12 plants	35 plants
Spinach	30 plants	60 plants
Squash	1–2 vines	5 vines
Tomato	2–4 plants	9 plants
Zucchini	1 plant	2 plants

Within the homestead garden, tomatoes are staked with bamboo poles. Poles are reused from year-to-year and have a wide variety of uses in creating trellises for climbing crops, acting as support stakes, and more.

Sunflowers are excellent pollinator attractors. They also provide support for climbing crops such as legumes and gourds and break compacted soils with their long taproots.

HOLISTIC PEST MANAGEMENT SYSTEM

Healthy soil structure promotes healthy plants. Healthy plants are equipped with better pest tolerance, and even resistance, through their thickly fortified cell walls and plant tissue, deep green leaves, protective wax coatings, and spiny hair follicles. It has been scientifically proven that insects prefer to feed on plant life with yellow-green foliage, soft or easily penetrable plant tissues, and no waxy or prickly protective barriers. A nutrient-dense soil habitat combined with permanent perennial vegetation that hosts beneficial insects will create a working ecosystem that naturally keeps "bad bugs" in check. This system takes time to achieve.

An Integrative Pest Management System (IPM) can be implemented on the homestead to reduce pressure from pests as the surrounding ecosystem establishes itself. Cultural, physical, biological, and chemical measures can be taken to reduce insect invasion. On an ecologically focused homestead, only the first three categories are used.

Cultural pest management methods revolve around environmental strategies when planting crops in order to minimize pest numbers. This includes polyculture, utilizing a crop rotation schedule, weed removal, and trap cropping. None of these actions are invasive nor do they interfere with the soil and its structure in any way.

When cultural efforts are not an option, the grower can turn to physical objects, such as row cover, fences, bird netting, motion sprinklers, and other barriers or deterrents. Physical pest management may be a more cost-effective option for small-scale growing spaces. The intention here is not to reduce the quantity of any given insect, but deter or physically reroute it.

The third facet of an IPM is the biological component. A permaculture homestead relies heavily on using beneficial insects and animals as a means to solving a challenge. For example, ladybugs are harmless to plants but feed heavily on aphids. If there is an aphid infestation, ladybugs can be introduced to control the population. Flowers and herbs which attract helpful insects can be interplanted within growing spaces to keep them returning and circulating. Other biological choices include hover flies, wasps, cats, poultry, bats, owls, and more. Look to chapter 5 (page 108) for more information on the contributions of differing animal species on the homestead.

Some growers resort to chemical usage, including synthetically manufactured sprays and powders. Though there are many naturally derived options on the market, such as neem oil and even diatomaceous earth, these applications cannot decipher between beneficial and harmful insects. Diatomaceous earth, for example, is a powder ground from fossilized aquatic organisms called diatoms. The fragments within the powder are sharp and abrade the skin of insects that it comes into contact with. But the powder kills not only slugs and snails; it also kills earthworms, butterfly larvae, and decomposition workers (e.g., pill bugs). The result is a loss of all inhabitants of the plants and surrounding soil.

Beneficial Insects Worth Attracting to Your Growing Spaces

Beneficial insects can offer multiple services, including the pollination of crops and flowers, predation of unwanted pests, and the parasitizing of undesirable insects. Planting hosts and food for these garden helpers can encourage a working ecosystem within growing spaces.

Pristine broccoli was grown using natural pest control methods, including the integration of garlic within the same raised bed to help repel cabbage moths.

HELPFUL INSECTS FOR THE GARDEN

Insect	Benefits	To Attract
Parasitic Wasps	Parasitize cutworms, cabbage worms, army worms, coddling moths.	Plant flowering herbs, such as dill, fennel, yarrow, and thyme.
Green Lacewings	Lacewing larvae consume aphids, corn worms, potato beetle larvae, mealybug eggs, lace bug eggs, spider mite eggs, and whitefly eggs.	Plant cosmos, yarrow, goldenrod, Queen Anne's lace, and daisies.
Praying Mantis	Consume moths, beetles, flies, and grasshoppers.	Incorporate marigold, dill, raspberry, and fennel.
Hover Flies	Hover fly larvae eat large quantities of aphids and mealybugs.	Plant daisies, marigolds, coreopsis, speedwell, veronica, lavender, asters, bugleweed, statice, sedum, dill, fennel, and feverfew.
Wolf Spiders	Wolf spiders feed on a variety of insects including earwigs, flies, crickets, ants, and grasshoppers.	Use a variety of organic mulch, such as leaves, straw, and grass clippings.
Ladybugs	Consume aphids in large quantities.	Plant dill, coriander, cilantro, thyme, oregano, and sweet alyssum.
Ground Beetles	Larvae consume slugs, thrips, weevils, silverfish, cutworms, caterpillars, tobacco (geranium) budworms, potato beetles, and asparagus beetles.	Incorporate stones, rocks, logs, and other places for these beetles to burrow under and hide.
Assassin Bugs	They prey on aphids, weevils, leafhoppers, cabbage loopers, and cabbage worm.	Intersperse daisies, dandelions, dill, fennel, goldenrod, Queen Anne's lace, marigolds, and tansy.
Soldier Beetles	Feed on bean beetles, potato beetles, caterpillars, and aphids.	Plant flowering herbs, such as parsley, dill, fennel, cilantro, and thyme.

CHAPTER *four*

INCORPORATING ANIMALS

To an outsider, it may look like my homestead is a collection of animals that I've procured over time for their cuddly characteristics. While I do certainly find my animals cute, I did not bring them to the farm without ensuring they could provide a unique contribution. Each and every critter here plays a role in our ecosystem that no other animal could. And, in many cases, our animals offer stacked functions—in other words, they provide more than one service to the homestead.

When looking for a horse for the farm, it was important to me that I could do more than just saddle up for an afternoon trail ride. I wanted a horse that could work with me to plow my fields. Ironically, I've become a no-till farmer and now have no need for plowing, but I still harness my horses for pulling harvest carts, a manure drag, and a seed spreader. I also learned to hot compost manure and use this nutrient-rich organic matter in my growing spaces. The horses are multifaceted in their purpose here, as is every other animal.

Incorporating an animal can often be a way of solving a problem. Rather than turn to herbicides for leafy weed control, I use geese and sheep. When I contracted Lyme disease along with two of my dogs one summer, I researched the best poultry solution for consuming ticks. One guinea fowl hen can eat hundreds per day. And, as mentioned in the next chapter, multispecies grazing offers parasite reduction.

Choosing the right animals for your homestead is key to a thriving ecosystem. The landscape, pasture footing, living spaces, feed, and water access all need to be taken into consideration to keep the animal healthy and happy—and to save the farmer excess money, energy, and time in trying to conform a livestock animal to a lifestyle that doesn't suit it naturally. In this chapter, I'll cover the most popular options for homesteaders, their basic requirements, and the benefits you can expect from each animal.

« I believe in handling and interacting with my animals as much as possible. This fosters a deeper relationship outside of work or grooming sessions.

New ducklings at the homestead begin their short field trips outdoors as soon as the weather is warm enough.

POULTRY

Fresh eggs come to mind first when most people think of the benefits of bringing home poultry. Rainbow egg collectors seek a wide variety of chicken breeds, and baking enthusiasts seek the moisture that only duck eggs can provide to their homemade goods. Eggs are wonderful, but chickens, ducks, geese, and guineas provide many other functions as well. And it's important to note they are all different species with their own unique needs.

Chickens

Chickens are known as the gateway animal into homestead animal keeping. Their small size makes them approachable for novice farmers, and they are easy to house in coops and runs. In addition to eggs and meat, chickens ingest insects while foraging and thus can help to reduce tick populations. As they free range, their droppings contribute fertilizer to the soil. In a pasture rotation system, chickens help to spread manure by scratching away while they look for insects.

Farm-fresh eggs are a bonus from our flock but not our primary purpose for keeping ducks and geese.

This same scratching behavior has earned chickens their notorious reputation for ruining vegetable and flower gardens. When left to their own devices, chickens will remove all grass or plant life and leave the soil bare. This can be avoided with portable runs (called chicken tractors) and limited access to pastures and green spaces.

Chicken is a favorite cuisine to both large and small predators. They will require the protection of a locked coop and run, especially at night. Inside the coop, chickens will need roosting bars for sleeping as they prefer to rest off of the ground. Their eggs are laid in small semi-private areas called nesting boxes, usually lined with pine shavings or straw. Their coop floor should offer bedding as well. Soiled chicken bedding can be composted and offered to the garden as a mulched fertilizer. (Note that bedding should be composted for a minimum of 120 days to successfully break down bacteria and pathogens that can contaminate crops and soil.) These birds do require supplemental feed inside their coop or run which includes chick feed (often medicated), grit, and scratch grains. A space for dust bathing helps them to feel and look their best.

Chickens are excellent backyard garbage disposals and are happy to feed on leftover kitchen scraps. Just make sure to avoid feeding any spoiled or rotted food items. Birds require constant access to clean drinking water.

Breeds for meat:	Sussex, Australorp, Orpington, Cornish Cross, Bresse, Leghorn, Freedom Ranger
Breeds for eggs:	Leghorn, Rhode Island Red, Sussex, Plymouth Rock, Ancona, Barnevelder
Breeds for foraging:	Jersey Giant, Ameraucana, Ancona, Andalusian, Buckeye, Egyptian Fayoumi, Hamburg, Welsummer

Ducks

Ducks are the poultry of choice for my homestead. They are friendly, imprint on their human family members, and are heartier than chickens in terms of health and cold tolerance. Ducks offer the homestead fresh eggs, meat, down feathers, fertilizer, and insect control. They are great additions to established gardens and vineyards for their love of eating snails and slugs. In pasture spaces, they are quick to root in manure piles for insects and snack on leafy vegetation. Note their bills do leave ruts within soft soils while foraging.

As with chickens, ducks require housing in a predator-proof run and coop to ensure safety. Their coop should be equipped with plenty of bedding as their droppings are wetter and messier than other poultry species. Ducks lay their eggs and sleep on the coop floor. Nesting buckets are appreciated but certainly not required. Soiled duck bedding can be applied immediately to the garden; however, fully compost it before using on vegetation with exposed eating surfaces, such as lettuce and greens.

Even when free ranging, ducks benefit from supplemental feed sprinkled with brewer's yeast for an added niacin boost. Grit should be offered, especially for birds who do not forage regularly. A constant supply of fresh water for drinking and bathing is a must-have for ducks as they are waterfowl. If a natural waterway is not available, kiddie pools can be filled and offered for swimming. At a minimum, ducks require a bucket of water for drinking as they need to blow the nares on their bills in order to breathe properly.

Breeds for meat:	Pekin, Cayuga, Ancona, Magpie, Buff Orpington, Rouen, Muscovy
Breeds for eggs:	Golden 300 Hybrid Layer, White Layer, Welsh Harlequin, Buff, Runner, Khaki Campbell
Breeds for foraging:	Ancona, Magpie, Runner, Khaki Campbell, Welsh Harlequin

When crops are young and tender in the spring, only a few ducks may be used within the garden at a time to reduce the threat of trampled seedlings.

Ducks forage the garden in summer after crops are well established and throughout the fall season. In winter, they have unlimited access as they help keep pests in control, fertilize the soil with their droppings, and loosen compacted areas with their bills.

DUCK BREED COMPARISON CHART

Various common duck breeds are listed below. This chart explores the differences in their temperament, weight class, egg color, egg frequency, and overall contributions to the homestead.

	Disposition	Weight Class	Purpose	Egg Color	Average Eggs/ Year	Exceptional Pest Forager
Pekin	Moderate	Heavy			200	
Crested Pekin	Nervous	Light			150	
Buff	Nervous	Middle			150–220	
Cayuga	Nervous	Heavy			100–150	X
Swedish	Calm	Middle			100–150	
Runner	Nervous	Light			200+	X
Ancona	Calm	Middle			210–280	X
Magpie	Nervous	Light			220–290	X
Khaki Campbell	Nervous	Light			250–340	X
Welsh Harlequin	Calm	Middle			240–330	X
Muscovy	Calm	Heavy			180	
Rouen	Moderate	Heavy			30–125	
Silver Appleyard	Calm	Middle			200–270	X
Saxony	Calm	Middle			190–240	X
Hookbill	Moderate	Light			100–225	

Geese

The emblem of the permaculture homestead, the goose has so many functions. In addition to eggs, meat, down feathers, and fertilizer, geese are excellent weeder animals who forage on pesky weeds and growth. They are unique in their watchdog abilities as territorial animals. Geese are very attuned to their surroundings and flock. This means they are quick to sound their calls to alert their flock mates and the homesteader of any abnormalities including predators, intruders, or unknown birds. Guardian geese can be raised from goslings alongside the chickens or ducks they are meant to protect. While they are great alarm bells, they are not physically able to defend themselves or their flock members from larger threats, such as fox and coyote.

Geese are paired animals and require another goose to be happy. They do not lay year-round and are seasonal egg producers, laying only from late winter through late spring. During breeding season, geese can become temporarily aggressive toward other flock members and even the farmer. This can be remedied by providing a separate housing space to these large birds while nesting. Both nests and sleeping take place on the ground. Breeds that are more docile, even during mating season, include the Large Dewlap Toulouse and Sebastapol.

Geese can fly but typically do not. They appreciate access to water—especially the Sebastapols, known for their long, curly feathers. A kiddie pool or water trough for bathing can suffice if no natural waterway is available. These birds require a bucket of water for drinking, just like ducks, in order to blow their nares and clean their beaks when drinking. Supplemental duck or waterfowl feed is needed to ensure all nutritional gaps are closed for the goose. Niacin by way of brewer's yeast is required for young goslings to ensure proper growth.

Geese do wonderfully on pasture and prefer to graze throughout their day. They often are referred to as lawn mowers for their grazing habits. Even though they are larger in size, geese still do require a coop and run as a safe haven from predators and harsh winter weather.

Breeds for meat:	Large Dewlap Toulouse, Buff, Cotton Patch, Embden, Pilgrim
Breeds for eggs:	African, Cotton Patch, Buff, Toulouse, Embden
Breeds for foraging:	Chinese, Cotton Patch, Toulouse, Embden, African, Pilgrim
Breeds for guarding:	Chinese, African, Roman Tufted, Embden

Geese are excellent alarm bells to any unusual behavior within their environment.

WHICH BREED OF GOOSE IS RIGHT FOR MY FARM?

Choosing which breed of goose is right for your farm or homestead can be overwhelming. But by deciding on what your primary intentions are with raising geese, you can certainly begin to narrow the choices. Answer simple yes-or-no questions and follow the path below to figure out which breed of goose is right for your homestead environment.

WHICH BREED OF GOOSE IS RIGHT FOR MY FARM?

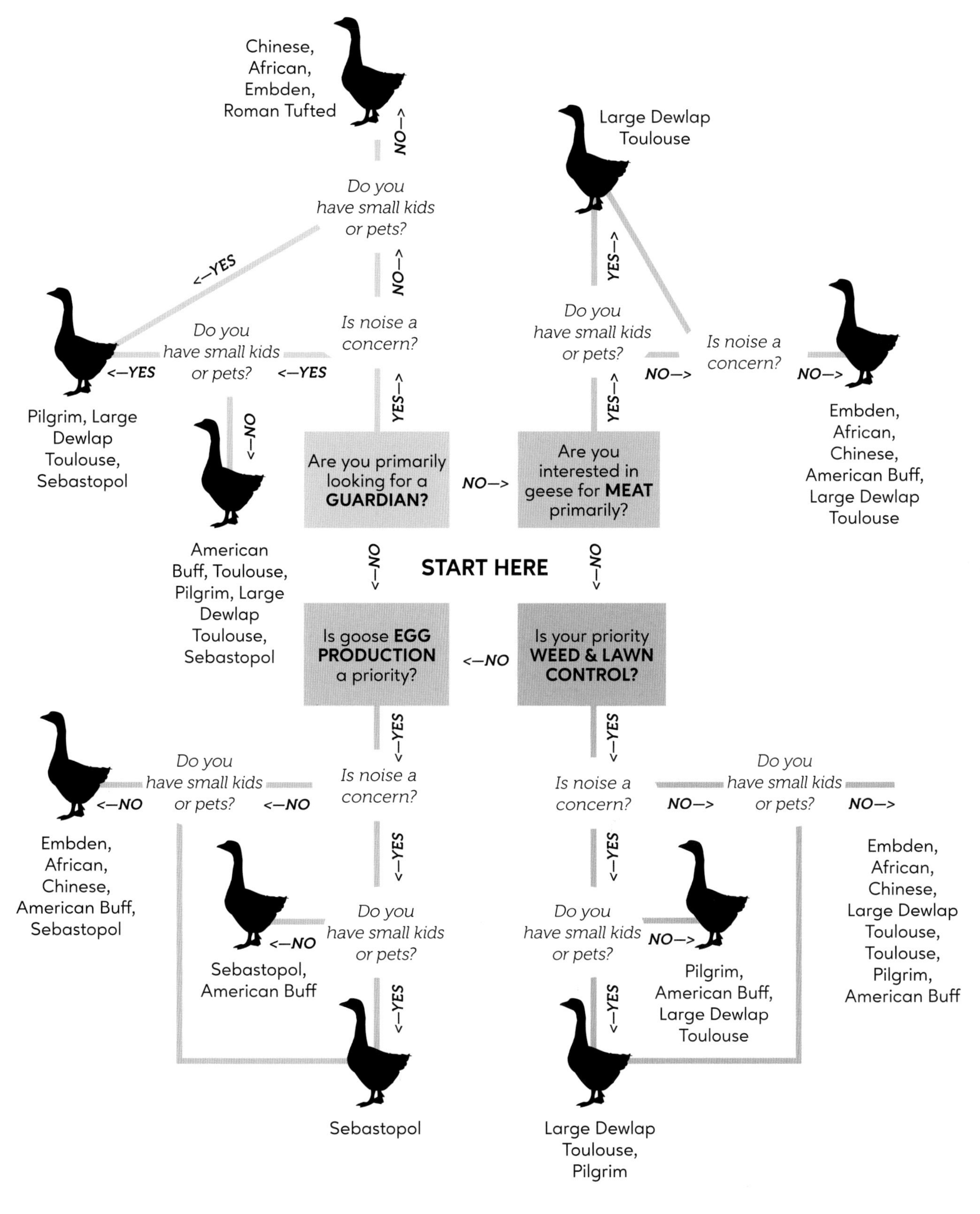

Turkeys

Turkeys have been raised on family farms for years as meat providers. But they also are valuable insect reducers as they forage for horn worms, hookworms, and more. As they forage, their droppings are left behind as fertilizer. Turkeys are seasonal egg layers, and they can produce up to 100 eggs per year, typically around March or April. A nest box is not required, but generally appreciated and preferred. Turkeys prefer to roost while sleeping and require an elevated roosting bar or tree in which to perch, especially during the night.

When given the ability to free range, turkeys graze on grasses and leaves, berries, seeds, and insects—especially slugs. Supplemental game feed with a high protein content is suggested along with scratch grains. Clean water access is required for drinking; however, these birds preen and maintain their feathers by dust bathing.

While turkeys are regal, and even graceful, in their appearance, they can have difficulty intermixing with other flock species. Some farmers have great success with mixing turkeys and chickens or ducks, but many struggle with aggressive turkeys who terrorize their smaller poultry.

Breeds for meat:	Bourbon Red, Blue Slate, Bronze, Narragansett, Royal Palm
Breeds for eggs:	Beltsville Small White, Bronze, Jersey Buff, Bourbon Red, Royal Palm, Narragansett
Breeds for foraging:	Narragansett, Blue Slate, Royal Palm, Midget White

Guinea Fowl

When you tell someone you're introducing guinea fowl to your farm, chances are they'll comment on how loud and disruptive these birds can be. Guineas are notorious alarm bells, sounding their shriek at any and all movement, abnormality, or visitor. Generally, they wander when allowed to free range, foraging far for insects and their signature prey—ticks. In addition to ingesting insects, snakes, weeds, grasses, and seeds while foraging, guineas should be fed a small, crumbled feed as a supplement. Plenty of clean drinking water is imperative for health.

Guineas prefer to sleep off of the ground as high as possible. If left to their own devices outdoors, they'll seek safety in tree branches. They are ground nesters, however, and find cover among branches and tall grasses for their eggs. They are seasonal layers, though their laying period is much longer than that of geese. They begin around late spring and finish the breeding season in early autumn.

These backyard birds are naturally territorial over their flock mates and land. Combine these characteristics with their noisy tendencies, and you have an excellent flock guardian. While guineas aren't physically capable of defending themselves and their flocks from larger predators, such as coyotes, fisher cats, and foxes, they are quick to sound their shriek at any threat, and they don't shy away from confrontation with rats or mice.

It is recommended that guinea fowl, ideally, be enclosed in a coop by nightfall for predator protection. Some folks have great luck allowing their birds to forage freely by day and watching them parade home at dusk. Other homesteaders have a terrible time gathering their guineas for nighttime lock-up. There are many training tips to entice guineas to return such as feeding only in the evenings, allowing only a select few birds to forage at a time, leaving the rest of the flock in the coop, and enclosing them for up to six weeks before letting them out at all. Regardless of whether they roam or remain in a run, guineas will need access to a dirt patch for dust bathing.

Guinea fowl are notorious for their loud alarm systems, alerting the farmer to any abnormality on the homestead. They favor ticks when foraging and can quickly reduce populations.

Breeds for meat:	French, Helmeted, Pearl Grey
Breeds for eggs:	Pearl Grey, White, Lavender
Breeds for foraging:	Helmeted, Pearl Grey, Lavender

The Sustainable Coop

Here are a few tips and tricks to make your coop a little eco-friendlier and more sustainable for your flock and your farm.

COMPOST BEDDING

When coop cleaning, cart the used bedding away to the compost heap. This nutrient-rich material should be fully broken down before applying to the garden. Soiled duck bedding may be applied immediately, but wait to harvest skinless vegetables, such as lettuce, for at least one week after application. Chicken manure should be composted for six to nine months to remove bacteria and pathogens.

REUSE THE WATER

When cleaning water buckets and pools, use the dirty water for the garden. Full of manure, this creates a liquid compost tea plants love. Apply to growing spaces after a harvest to avoid contact with edible surfaces.

INCORPORATE HERBS

Herbs are a great insect repellent. I hang bunches from my coop ceiling and also sprinkle herbs (both dried and fresh) in my nest buckets for my ducks and geese. Herbs that are safe for poultry include mint, lemon balm, lavender, anise hyssop, oregano, thyme, sage, basil, and rosemary.

When herbs are abundant, large bundles are harvested and hung from the coop ceiling. Many herbs such as anise hyssop help deter flies.

RECYCLE NESTING BOX BEDDING FOR RUN BEDDING

My girls use nesting buckets for laying their eggs. After the eggs are laid, they leave the buckets. This means the bedding is fairly clean, aside from the occasional bit of muck. I prefer to replace the pine shavings in my nesting buckets about once per week. I dump the old nesting bedding outside in the enclosed duck and goose run as fresh footing. This gives the bedding a second use.

CAYENNE PEPPER

An old wives' tail states that sprinkling a bit of powdered cayenne pepper on top of poultry feed acts as a dewormer. Turns out, it's true! Capsaicin, a compound found in cayenne pepper, was found by scientists to stimulate laying performance in poultry, and even act as a possible alternative to antibiotics and dewormers. Poultry do not have the oral receptors to taste the spicy flavor of cayenne, so it is of no bother when used as a garnish to their feed. This powerful spice also enhances intestinal and pancreatic activity. I use this with my own flock regularly, and I can attest to all of these points personally.

REDUCE FOOD WASTE

Poultry are notorious for walking through their food and water dishes, tipping them or depositing droppings in them as they go. To prevent soiled feed, I built a raised trough out of a few lumber scraps for my poultry. With their rations off the ground, they can't dirty their feed. If you try the same, be sure to use untreated wood.

CEDAR SHAVINGS

Cedar shavings should never be used as bedding in livestock living quarters as it causes respiratory distress. However, it's an excellent fly repellent in coops. Use it in place of standard pine shavings to line nesting boxes or sprinkle throughout the coop floor.

CHICKEN, DUCK, OR GUINEA TRACTORS

A poultry tractor is a portable coop or run to contain the birds and keep them safe from predators while the flock forages. These structures are often very simple in terms of design and complexity; a solid perimeter of wood creates a base and inside resides an open floor allowing for foraging on vegetation and insects. Cattle paneling, PVC piping, or aluminum prefabricated runs typically create the walls and roofing frame. Plastic blanketing, a tarp, or polycarbonate roofing panels are veiled on top to provide shade and protection from the elements. Wheels are affixed to the exterior of the baseboard allowing for mobility. As the flock grazes, they consume all of the available forage within the provided real estate. Then the structure is moved several feet to a new plot of vegetation to restart the grazing cycle, as with rotational grazing. The droppings of the birds fertilize the soil as they go.

Sheep are well-suited additions to farms with grassy brush and weeds, as this is their preferred forage.

SMALL HOOFSTOCK

I've spoken with so many homesteaders who enter the world of animal keeping with poultry. They get a taste for fresh eggs and then the world of livestock becomes appealing. Surely if fresh eggs taste better and are more nutritious than store-bought, what about home-raised dairy or meat? Small hoofstock animals are a natural progression as they are smaller, easier to house and fence, and often easier to physically manage than larger stock. The beauty of introducing hoofstock is that they are often multipurpose in their contribution to the homestead. A sheep kept for meat can also offer dairy for cheesemaking and, of course, fiber. Goats can also provide similar stacked functions.

Sheep

We brought sheep to the farm for a myriad of reasons. The first being their preference for grass. Sheep don't need supplemental feed if healthy pastures are in place. Hay in the off-season, loose minerals specifically formulated with little-to-no added copper, along with plenty of clean drinking water are all that's necessary to keep them happy and healthy. Our brushy and overgrown areas aren't really brush at all . . . just really tall grass. Sheep help us maintain and work these spaces, increase our carbon absorption, and keep a closed loop grazing system in place with our horses. All of these concepts mean we can keep sheep well fed and healthy without adding any required grain. And in return, they feed the soil with their droppings as they forage.

In addition to being well suited to our landscape and creating a perfect grazing system with our horses, sheep give us wool. Most sheep require shearing annually in order to prevent matting, overheating, fly-strike, and an onslaught of other health conditions. A professional shearer can be hired to quickly shave the animal without causing knicks or discomfort. Most will provide hoof trimming services as well, though the sheep will likely require much more trimming than once annually based on the landscape they are exposed to.

Yarn is only one of the ways sheep can contribute to the homestead.

A Note on Mulesing

There is some controversy about animal cruelty with regard to shearing. This is largely due to confusion and association with the practice of mulesing. This is a completely separate and largely illegal procedure thought to reduce the effects of fly-strike. Fly-strike is a painful (and sometimes fatal) condition where flies lay their eggs on a host animal. Once the larvae hatches, they feed on the skin of their host underneath the wool. For sheep this occurs most commonly on the rear of the animal. Thought to reduce infection, some farmers practice mulesing. Mulesing is the removal of wool and its underlying skin around the breech (buttocks) of the sheep with sharp shears. The removed skin surrounding the tail scars over as it heals leaving a smooth surface clear of wool, reducing the opportunity for flies to lay their eggs and cause harm to the sheep.

Mulesing is a painful procedure for the animal and is illegal in most countries. To date, Australia is the only country which allows this practice. Unfortunately, 70 percent of the world's merino wool comes from Australia, and merino wool is the most common fiber used in wool goods. If you wish to show support for mulesing-free fiber farmers, ensure to purchase certified mulesing-free wool products—or make your own with yarn from local farms.

Wool is most commonly associated with yarn or clothing. It keeps sheep (and people in the form of knitted wearables) warm in the cold months and cool in the hot season. Wool can go longer between washes than many clothing fiber choices due to its natural sweat wicking, anti-microbial, and stain-resisting qualities. And when not blended with any additional plastic fibers or unnatural dyes, that unwanted sweater or outgrown sock can be added to a compost heap. Wool is fully biodegradable and a renewable resource. Some homesteaders use wool as a natural mulch on their gardens as it protects the soil between their crops while it slowly decomposes over time.

When not accompanied by the protection of a Livestock Guardian Dog, sheep should return to a barn stall or enclosure by night. Coyote and large game predators have no problem preying on sheep for a meal. Clean bedding in the form of dropped hay from feeders or straw is always appreciated for a soft place to rest at the end of the day.

Of course, some folks incorporate sheep into their farm portfolios for a meat or dairy source. Folks who have sensitivities to cow's milk may find sheep's milk to be easier to digest. Many cheeses can be made with sheep's milk, including Manchego, Roquefort, pecorino, and feta.

Breeds for milk:	East Friesian, Lacaune, Icelandic
Breeds for meat:	Romney, Hampshire, Suffolk, Dorper, Babydoll Southdown, Texel, Dorset, Shropshire, Icelandic
Breeds for fiber:	Merino, Debouillet, Rambouillet, Cormo, Teeswater, Bluefaced Leicester, Corriedale, Romney, Icelandic
Breeds for foraging:	Romney, Shetland, Suffolk

Romney sheep contribute to the farm with their pasture rotation contribution, droppings for the soil, and their wool for yarn.

Goats

Goats are very common small hoofstock additions to both large- and small-scale homesteads. Their generally compact size and voracious appetites for weeds and woody brush make them useful additions. A small number of goats can be a wonderful contribution to a permaculture environment, however, when not cared for properly or when overpopulated, they can cause more harm to the farm than good by overgrazing, damaging soil structure, and eating perennial fruit or nut trees.

It is true that they are happy to forage on poison ivy, and they seem to thrive on less-than-ideal pasture grazing spaces. Unknowing homesteaders are baffled when they introduce goats to a pasture of grass and wonder why they are browsing for weeds first and grass second; it's because these animals have an appetite that's unique. When not in the field, goats can be fed free choice hay and minerals. Of course, plenty of access to clean water also is necessary. Grain rations may be added to the diet based on veterinary recommendations to ensure all nutritional needs are satisfied. Free choice minerals should also be provided to any goat to maintain healthy body functions.

The inquisitive, playful personalities of goats have earned them the reputation as escape artists. They feel compelled to climb fencing to discover what's on the other side, climb housing structures, and to explore feed/supply closets (which should be secured and locked). Strong, sturdy, tall fences are required to keep these animals contained and safe from roadways and threatening situations. Play equipment, such as logs, tires, and repurposed and modified swing sets, can all cure boredom and encourage exercise—especially for goat kids. Large rocks help to keep the hooves naturally filed down and reduce trimming sessions. When unaccompanied by a Livestock Guardian Dog, goats should be kept in a predator-proof enclosure by night for safety from large predators, snow, rain, and excessive wind.

Goats create a closed loop system when used in a multispecies grazing rotation (see page 116). They forage on vegetation other animals will leave behind and ingest and terminate many parasites that affect horses and cows. Their droppings return nutrients to the soil as they go, and soiled bedding can be collected and composted. Goats are commonly used on homesteads for their milk, which can be used for everything from chèvre to soap. They can also be harvested as a protein source. There are several breeds that offer fiber such as Angora, Pashmina, and Cashmere. Rather than one annual shearing, these fiber goats are shaved twice yearly for a fiber harvest.

Breeds for milk:	Alpine, Saanen, Nigerian Dwarf, Nubian, American LaMancha, Boer, Oberhasli
Breeds for meat:	Boer, Nubian, Pygmy, Spanish, Kiko, Tennessee Fainting, Angora, Don
Breeds for fiber:	Angora, Altai Mountain, Hexi Cashmere, Pashmina, Don
Breeds for brush clearing:	Alpine, Boer, Pygmy, Angora, Hexi Cashmere

Pigs

Pigs are not considered a permaculture hoofstock animal. They do not graze on forage that regrows nor keep soil structures intact. Rather they root and overturn stumps and logs that would otherwise break down and return nutrients to the soil. Roots that once held carbon are pulled up, releasing those elements back into the atmosphere. They can, however, be used as a tool in a permaculture homestead.

When left to their own devices, pigs will do what their innate behavior instructs them to: root, wallow, wander, breed, and play. They are excellent consumers of food scraps, so no kitchen remnants go to waste. In addition to the deforestation of wooded areas, pigs offer the farm bacon and pork, plus manure that can be composted into an organic growing substrate. They are not part of a rotational grazing plan though, as their destructive rooting will quickly ruin a pasture space.

Some farmers choose to use rings in the snout of their pigs to discourage rooting. But homesteaders can also create their own pasture just for pigs subdivided into smaller sections. Pigs are rotated through this divided paddock after fodder has grown specifically to meet their dietary needs including brassicas, peas, beets, corn, and clover. Hay can be used to feed pigs as supplemental forage—plus excess hay (or straw) makes excellent bedding for nesting. Fresh clean water should be provided for drinking. Of course, a wet, muddy puddle is appreciated for cooling themselves off in the hot summer heat.

The larger the pig, the stronger the fence needs to be. Large hogs can grow to massive proportions and have the strength to match. Heavy duty fence posts, barbed wire, electric fencing, and woven wire should keep them contained. A strong shelter with plenty of ventilation is suggested for protection from harsh weather conditions.

Breeds for meat:	Yorkshire, Duroc, Berkshire, Meishan, Landrace, Chester, Hampshire, Pietrain, Tamworth
Breeds for brush clearing:	Tamworth, Berkshire, Mangalitsa, Red Wattle, KuneKune, Chester White, Hampshire

Sustainable Housing

Here are a few tips and tricks to make your small hoofstock animal housing eco-friendlier and more sustainable.

COMPOST THE BEDDING

Fresh pig, sheep, and goat manure contains a multitude of bacteria and pathogens including *E. coli, Salmonella*, and parasitic worms. While it should never be applied when fresh to growing spaces laden with crops for consumption, it can certainly be composted. When cleaning the shelter structure, cart the used bedding away to the compost heap. This nutrient-rich material should be fully broken down before applying to the garden. This process usually takes about four months using the hot composting method for these manure types.

FLY PREDATORS

Flies can reproduce quickly in coops, stalls, and animal housing structures. Sticky traps and bait bags are certainly useful, but they work to solve the problem well after it has started. Fly predators (see more in chapter 5) are small beneficial wasps that lay their eggs and feed on the larvae of various fly species. These can be ordered from many online retailers and are typically peppered throughout living spaces on a routine basis.

CAYENNE PEPPER

Just as with poultry, small hoofstock benefit from a small dose of cayenne pepper. A sprinkle of the powdered spice on a daily feed ration helps to reduce parasite numbers, stimulate circulation, and support intestinal health.

GRAY WATER

Gray water is water that is slightly dirty from the mouths of sheep, goats, and pigs in their drinking buckets or troughs. When cleaning water buckets, save this water and reuse to irrigate crops or water flowerbeds. If the water contains salt or mineral from the animal, use this on heartier items such as garlic and corn.

ACORNS

Oak trees provide many uses to the homestead, one of which is acorns. If allowed to fall freely and scatter, these small nuts can be tedious to collect, and they are toxic to horse and cattle. Luckily, letting pigs feed on acorns is something that has been done for centuries—long enough that they say, "a pig finished on acorns will add a nutty taste to the meat."

The practice called *pannage* is the act of allowing pigs to graze an area to clear the acorns before allowing other stock to pass through. This prevents harmful ingestion to other livestock. Enzymes in the pig's mouth break down the tannins that cause many issues in other animals. That said, there is conflicting information about whether or not pigs should ingest acorns. Many farmers or homesteaders believe acorn ingestion to be harmful and cause a host of health issues including fetal abortions in sows. My best advice is to discuss all forage and dietary opportunities with your veterinarian. They will be able to best guide you in the right direction based on your animals' current state of health.

LARGE HOOFSTOCK

Most folks believe that larger animals mean larger feed bills, higher quantities of manure to clean up after, larger veterinary bills, and bigger housing structures. On a permaculture homestead, much of this is still true, but large hoofstock often do well on pasture alone to meet the majority, if not all, of their nutritional requirements. When given the opportunity to rotationally graze in clean, hygienic environments, parasite loads reduce as do bacterial infections. To me, animals who require more grazeable forage simply translate to more land, which can absorb carbon into its root systems as it regrows. If stock densities are kept in proportion to the land that is offered, these animals can be incredibly helpful partners on the homestead.

Horses

Clearly I am biased when it comes to horse keeping. These animals have become a valuable player in my homestead. They provide easily compostable manure that returns loads of nutrients to pastures, garden beds, and other growing spaces. They contribute to the multispecies grazing circuit, ingesting and killing off many parasites that infect sheep and goats. I use my horses for driving (pulling) and they help spread my fields with manure where the ducks did not break up all animal droppings from previous rotation cycles. I no longer till my soil to maintain optimal structure, so my drags are turned teeth up and attached to the horse via a single-tree harness. Any manure underneath is spread without abrading the soil's surface. And lastly, my horses are my companion animals; I ride on trails when the weather allows.

Draft horses produce manure for compost, contribute to the multispecies rotational grazing system, provide pulling power, and are fun to ride.

Finnegan, my Clydesdale gelding, knows the way home after a short work session in the lower field.

Horses in healthy condition do well on pasture. Depending on your geographical location, hay should be offered constantly as a supplement if enough fresh grass and vegetation is not available. I personally no longer feed my horses grain rations for many reasons both economical and environmental; instead, I offer a pelleted vitamin, a protein/fat supplement, and continuous access to salt blocks. A steady supply of fresh, clean drinking water is of the utmost importance for horse health and to avoid colic, a condition similar to bloat in sheep and goats.

At my farm, I converted a large three-stall car garage into a stable. Its walls are deep enough that the horses can safely escape wind, rain, snow, and sleet as they please. They do have constant in-out access to one of their pastures to encourage exercise and natural movement. I have found pelleted pine bedding to be the best bedding personally; it's absorbent and quick to break down in the compost pile. As for stall gates, heavy duty 10- × 2-inch (25- × 5-cm) boards were used to create custom builds and the same lumber is screwed into 2- × 4-inch (5- × 10-cm) studs for stall wall siding. Many farms use railroad ties, especially as corner posts. No matter what you use, posts should have the ability to withstand heavy weight and pressure. I have installed polybraid electric wire around pasture perimeters, and it is all connected to a solar charger to keep my drafts in.

For me, learning to drive a horse and capitalize on horsepower for work required great effort as not many farmers still practice this skill set in my location. I read books and articles, watched videos, and spoke to a few people who had either dabbled in it themselves or knew someone who had. My best teacher was Finnegan, my rescued Amish Clydesdale, as he had performed under harness for years. Watching, reading, and understanding the cues that worked and what didn't were of the largest help to me. I believe this time spent together between farmer and horse, accomplishing a task, is a bonding experience.

There's not a drop of gasoline involved, nor time spent fixing an engine. But horses do require regular hoof care, which includes trimming at a minimum. The decision to provide shoes for your horse is largely dictated by the footing the horse is exposed to. Soft and wet landscapes will cause more risk of an abscess to a horse with no shoes. Leather pads can be shaped and installed between the sole of the hoof and the shoe itself to provide a barrier of protection from wetlands.

As hooves continue to grow throughout the lifetime of a horse, so do their teeth. A horse dentist will come generally once per year to float (file down) the teeth of the horse. If left neglected, hooks and ramps can turn into sharp protrusions that can pierce the horse's palette and gums. Furthermore, horses require flat teeth to grind their food fully before swallowing. A horse's nutrition intake is related directly to how well their teeth grind their forage.

Breeds for foraging:	By nature, the digestive system of a horse is designed to eat forage around the clock. For this reason, any horse (unless they are sick or have specific nutritional needs or sensitivities) will work in a rotational grazing circuit.
Breeds for work:	Clydesdale, Halflinger, Morgan, Percheron, Belgian, Shire, Suffolk

Cows

Cows are a large livestock option for those looking for a sizable food source. Unlike horses, cows offer dairy and beef. They also fulfill a role in the cycle of multispecies grazing while providing manure that can be composted and used in various growing spaces. Larger breeds and oxen can be used as plowing animals and trained to work under a yoke (wooden plowing collar). I've also seen the rare occasion of a cow broke to give saddle rides!

Backyard dairy cows require milking twice daily post calf-delivery. Though some consider this task tedious, others find it to be a source of bonding between human and animal. Stanchions or milk stands can be home-crafted to demobilize the animal while milking. For those with lactose sensitivities, A2/A2 cows may produce milk that is more easily digestible as it only contains one type of protein: A2 beta-casein. Most grocery store milk products contain A1/A2 beta-casein proteins, which research has shown may be digested differently by the human body than milk containing A2 proteins only.

In the pasture, many cows are happy to graze on grass and leafy forages, such as alfalfa and clover. Hay should be offered constantly as a supplement, especially during the colder months where grass is not readily available. Many homesteads feed hay and pasture alone, however, a veterinary professional or animal nutritionist may suggest feeding corn silage, oats, or other grains. Salt is a highly important nutrient in the cow's diet as it helps the body to regulate water intake along with other vital organ functions.

Farms that are fencing in cattle should install strong and sturdy fence posts and gate systems accompanied by electric wire, cattle panels, or barbed wire. A shelter such as a barn or stable provides shade in the hot summer months and also a place to seek refuge from harsh winter weather if desired. A separate milking house or pen is appreciated by many home-milkers for the cover it provides in wet, rainy seasons.

Depending on the terrain, cows often do not need hoof trimmings regularly. Cows have fewer teeth than equines; their mouths are equipped with teeth on the bottom and a dental pad that helps in the grinding of forage on the top. As the cows age, lower teeth can become problematic and floating, filing, and other dental treatments may be required.

Breeds for milk:	Jersey, Guernsey, Brown Swiss, Ayrshire, Holstein, Milking Shorthorn
Breeds for beef:	Holstein, Angus, Shorthorn, Hereford, Dexter, Brahman, Beefmaster, Piedmontese, Herefordshire, Gelbvieh, Limousin, Belted Galloway, Longhorn, Welsh black
Breeds for foraging:	Hereford, Beefmaster, Limousin, Simmental, Angus, Charolais

Donkeys and Mules

Donkeys and mules, though both equids, have their own nutritional needs, behavioral traits, and health requirements. Donkeys and mules both eat less per pound of body weight than a horse and, when fed straw, hay, and a variety of grass and weeds from foraging, do very well without added grain. Only the hardest working mule or donkey will require a small grain ration according to veterinary instruction. Orchard grasses and timothy hays are excellent choices for roughage; no legumes such as alfalfa are necessary. Naturally, plenty of clean, accessible drinking water is required.

Donkeys and mules can be trained to work under harness, pull loads, and offer saddle rides. Their boxy-shaped hooves give them a sure-footedness that horses do not have. In addition, many homesteaders and farmers feel their donkeys and mules possess stronger hooves than those of horses as they are less prone to chipping, cracking, and breaking.

Mules, being the hybrid of a donkey and horse, are said to be quite sound and healthy. Their stockier legs are stronger in proportion to their body framework and less prone to injury and lameness. Some veterinarians feel that their hybrid organs function better compared to horses, and that mules possess overall healthier body functioning. It's not uncommon to see a mule retire at the age of thirty from work under harness or under saddle and live into its forties. Perhaps this can be attributed to the theory of self-preservation; both mules and donkeys are not afraid to work, but rather, constantly assess and work in moderation. They pace themselves according to their capability when it comes to work, are able to withstand extreme heat, and even slow their intake of drinking water when supply is limited to simply replace lost body fluids. For this, they rarely succumb to injury or death by being overworked, from heat stroke, or from dehydration.

As with any livestock animal, proper fencing and shelters are recommended. A barn, stable or lean-to will allow the animal a place to rest in the shade or otherwise find relief from harsh weather conditions. Their soiled bedding and manure can be composted for use in growing spaces around the homestead. Mules and donkeys also have a place in the multispecies rotational grazing circuit, returning nutrients by way of their manure as they forage. Just as with horses and cattle, the digestive system of donkeys and mules are not hosts to parasites that affect sheep and goats. Therefore, they can assist in breaking up the parasite life cycle.

There is a common misconception that donkeys make excellent livestock guardians. This is false; donkeys are equines and are prey within the food chain. They cannot defend themselves or their barnyard family members against predators, especially large game such as mountain lion, wolves, and grizzly bears. Though they are wary of strangers and environmental abnormalities, they are not true farm guardians and should not be expected to fill this role.

Breeds for foraging:	Standard, Miniature, Mammoth, Mules
Breeds for work:	Grand Noir du Berry, Standard, Mammoth, Mules

The Sustainable Stable

Large animals require larger quantities of hay and bedding, and they need larger living quarters as well. With more needs, there are opportunities for more waste. However, there are many ways to create a more sustainable stable.

ALTERNATIVE BEDDING

Pine shavings are commonly used as bedding in most stables. However, both pine pelleted bedding and corn cob bedding are great alternatives that decompose much faster than pine shavings in a compost setting. Corn cob bedding is the product of ground corn cobs, combined with steam and water. Pine pelleted bedding begins with high-heat-treated pine shavings. The dust is then extracted while softened fibers are shaped into pellets. Both bedding options are highly absorbent and biodegradable.

FLY SPRAY RECIPE

Our equine and cattle friends can easily become the victim of biting flies throughout the warmer months of the year. A great homemade fly spray that keeps the horses cool and comfortable, without using toxic chemicals, is below. Be sure to always conduct a patch test on any animal to detect any allergies or sensitives before coating the entire body.

Yield: 4¼ cups (about 1 L)

Ingredients

- 4 cups (940 ml) apple cider vinegar
- ⅛ cup (30 ml) olive oil
- 2 tablespoons (30 ml) MTG (see note)
- 2 tablespoons (30 ml) unscented dish soap
- 20 drops lavender essential oil
- 20 drops rosemary essential oil
- 20 drops peppermint essential oil

Directions

1. Mix all the ingredients and pour into an unused weed sprayer or clean spray bottle. (If your container is large enough, you can add everything directly to the sprayer.)
2. Shake before use, and spray away on your horses! For best results, spray on once in the morning and again in the late afternoon. Also best applied after a sweaty workout.

MTG is an equine grooming product called Mane•Tail•Groom. It's available at most agricultural and feed stores and also on Amazon.com

REUSE HAY WASTE

No homesteader likes to see quality hay dropped to the ground and left as waste. For the exact same reason that fresh hay should never be used as mulch in a growing space—because it's loaded with grass seed—you can make use of hay scraps by spreading it back into pastures or traffic areas. It provides a barrier to exposed earth from solarization and nutrient loss, and also prevents erosion. Those small grass seeds will sprout in areas where they're actually welcome!

GOLDFISH IN THE WATER TROUGH

Large stock tanks are the perfect breeding location for mosquitos. Though the practice can be controversial, goldfish are excellent at keeping unwanted mosquito larvae out of troughs and are happy to forage on bits of hay and grass that drop from the horses' mouths when drinking. I've been using goldfish in my troughs for years—no bubbler required! Algae, which grows naturally on the sides of the stock tanks, produces oxygen during the process of photosynthesis. The algae actually produce more oxygen than they can consume, and this provides an adequate supply for the fish.

The stock tanks still need cleaning as fish do not help with algae and dirt. To ensure my fish receive enough oxygen, but to ensure a clean trough for my horses, I find scrubbing the trough once every seven to ten days to be adequate. Scoop them out and set aside in a bucket of water with the same temperature while cleaning. Clean the trough with soap and water and return the fish. The horses do not ingest the goldfish as they swim to the bottom when the horses approach. They overwinter well, especially with the help of a stock tank heater if available. (Note that heaters should never be used indoors.) The manure that fish create also do not cause the horses any health issues—after all, fish are naturally occurring in waterways.

Spent hay can be collected and used to reseed pastures. This covering of loose hay also acts as a mulch preventing solarization, erosions, and water runoff.

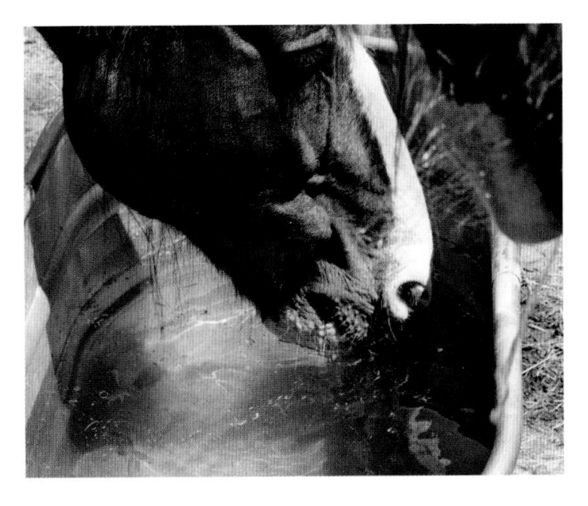

Two or three goldfish are employed per each of our 100-gallon (378 L) stock tanks to help reduce mosquito populations.

Livestock Guardian Dogs protect our stock from coyotes, fox, raccoons, opossums, mountain lions, bobcats, and more.

Livestock Guardian Dogs desire to stay with their stock even in the most extreme conditions.

OTHER CONTRIBUTING CRITTERS

Poultry and hoofstock aside, many farms and homesteads complement their barnyard families by raising barn cats as mousers, partnering with Livestock Guardian Dogs for protection, or keeping honey bees for pollination and honey.

Livestock Guardian Dogs

All Livestock Guardian Dogs are farm dogs, but not all farm dogs are livestock guardians. Livestock Guardian Dogs belong to a specific class of dog breeds that have been bred over centuries for their independence, problem-solving skills, ability to defend their stock, and for a delicate manner with their charges. These dogs are alert and incredibly astute to their surroundings and the well-being of the animals they are meant to protect. Though fierce and willing to give up their life while defending their stock, they are not aggressive to their handlers. They do not immediately charge at or attack intruders either; rather, there is a progression of behavior to serve as a series of warnings. First, the head and tail are raised as a sign of attention. Second, the dog will bark. Third, the animal will run toward and charge at the threat, while barking, in an effort to scare away the intruder. If all of these efforts are of no consequence, the dog will physically defend its home and barnyard family members if needed.

These dogs are partners and require significant consistency when training. They are much different than domesticated house dog breeds and respect is earned from the farmer; it's not automatically offered. Typically, guardians are raised from a young age alongside the stock they are meant to protect. They generally live with, eat with, and are constantly surrounded by their charges in order to see them as their own.

Here at my homestead, I use Livestock Guardian Dogs to hold the line between where my farm property ends and where nature begins. I believe this approach allows us to farm alongside nature—something we could not do without losing crops, poultry, and even cats before we had our dogs. I have chosen not to use traps and guns as a means to keep hungry predators away. Instead, I use an Anatolian Shepherd and a Great Pyrenees to communicate our territory. Though we have only six acres (2.4 ha) of land, we have a heavy predator load, including bobcats, black bear, coyote, fisher cats, fox, opossum, raccoon, and even the occasional mountain lion. The number of guardian dogs to employ is not determined by the number of stock you have. It is determined by the predator pressure.

Livestock Guardian Dog breeds:	Anatolian Shepherd, Great Pyrenees, Maremma, Akbash, Turkish Kangal, Karakachan, Kuvasz, Komondor

Honey Bees

There was a point in time when I advocated strongly for helping honey bees by building their populations through becoming a beekeeper. While I still feel this is important, I do believe that contributing to the native pollinator population is just as critical. Honey bees were introduced to the United States in the 1600s. Before that, farmers depended solely on native pollinators to pollinate crops. Feeding honey bees as well as natives, such as mason bees, bumblebees, carpenter bees, and more, is simple and extremely beneficial to crop yields. All you have to do is plant a variety of flowers throughout the homestead. A few favorites are coneflower, bee balm, mint, wild bergamot, anise hyssop, lavender, salvia, sunflowers, rudbeckia, hollyhocks, and poppies. Fruiting trees, such as apple, pear, plum, peach, cherry, almond, etc., also are attractive to these foragers. These perennial plants will return every year, contributing to soil structure, pulling nutrients into the soil, and loosening layers.

In addition to raising honey bees in homestead apiaries, constructing bug hotels or mason bee homes is a great way to house natives. For example, a mason bee house consists of hollow tubes, such as 4-inch (10-cm) cuts of bamboo poles, placed inside a wooden box or aluminum can. Mason bees are solitary creatures and do not live in colonies. They lay their eggs inside the tubes, cover with mud, and move on their way.

As a beekeeper, harvesting honey was never my primary objective—it was increasing my produce production. Still, honeycomb is an incredibly useful resource on the homestead for making salves, candles, and balms. Honey certainly comes with its own benefits, such as for selling, eating, and using as a natural sweetener and for its medicinal properties.

Getting back to the benefit I love most: a larger quantity of bees, be they native or honey, means more foragers looking for pollen and nectar. If you have blossoms on your plants with little fruit production, consider interplanting flowers. Pollinators will find the flowers and ultimately find the fruit and vegetable blossoms as well.

CHAPTER *five*

DESIGNING A PASTURE

Making the decision to incorporate poultry and livestock into the homestead portfolio requires planning the farm and equipping it with the right structures. Coops, barns, stables, and shelters should be in place before the animals arrive. A pasture system with sturdy fencing and gate infrastructure in place is just as important. Even better is a grazeable pasture arrangement that accommodates a rotation schedule.

The entire purpose of a pasture is to contain livestock while they feed. The forage they ingest from these spaces should meet the majority, if not all, of the species' nutritional needs with regard to energy and crude protein. It often is overlooked that animals are a major contributor to soil structure and health from their grazing habits. When the animal-soil nutrient cycle is allowed to take place, nutrients by way of animal manure are fed to the soil, and the soil can provide more nutrient-dense forage for the animal to graze. In addition, forage that is mowed through animal grazing will regrow. As this period of regeneration takes place, the plant undergoes photosynthesis. This process absorbs carbon from the atmosphere. The more a plant or forage crop is trimmed and allowed to regrow, the more carbon the plant can absorb. This carbon is then leached into the soil and safely removed from the atmosphere.

The key to capitalizing on photosynthesis, carbon absorption, fast forage regrowth, and nutrient-dense vegetation is rotating the animal through various grazing spaces. If animals are not regularly rotated, forage becomes overgrazed, is slower to grow back, and loses nutritional density, and both the animal and the soil health suffers. Muddy footing ensues, parasites abound, and costs for hay supplementation rise. Stubborn weeds remain in difficult growing conditions and, often unpalatable to most livestock, they are left to reproduce at a rapid rate. Soon a lush forage space can become a weed-ridden mess.

« Our draft horses are used for activities that don't disrupt soil layers. We do not till or plow. Rather, they help us bring in the harvest, spread manure, and seed our pastures.

PASTURE-ANIMAL COMPATIBILITY

Not all grazing spaces are ideal for all species. Even if a grazing space is outfitted intentionally with cover crop forage, the native growth, any standing water, the footing, and the terrain are all to be taken into consideration before fencing is installed and before the integration of animals. Goats, for example, are browsers and prefer to eat with the head raised above shoulder height. A goat would not necessarily be content on a pasture consisting of solely field grasses. Sheep, however, would much prefer these conditions. A pasture space containing brambles, weeds, barberry, and poison ivy would make the perfect forage for goats. Sheep would not be nearly as excited.

Horses prefer to eat the very top of the grass blade.

Pigs are very versatile in their forage preferences and often are brought to homesteads as a way to clear woodlands for the creation of silvopastures. A silvopasture is a grassy forage space on the forest floor integrated with trees throughout. This is an approach to forest management that works to clear mid- and understory growth in favor of the overstory. Swine are helpful for this because they are perfectly content to graze and root in almost any setting and work quickly to clear brambles, overturn stumps, and loosen root systems. In doing so they will leave ruts behind along with overturned soil. Though this is a natural method of land clearing and fertilizing, keep in mind this action heavily disrupts existing soil structure, releases sequestered carbon and nitrogen back into the atmosphere and removes perennial roots, which is why many ecological farmers do not consider this a permaculture practice. After the mid- and understories have been cleared, grass and other forage typically is sown in its place, making way for cattle, sheep, and other stock to graze under the tree canopy.

In addition to natural forage, footing also is important. When I brought home horses, I had no intention of making them wear shoes. Of course, I understood the importance of proper farrier care and budgeted accordingly for hoof trims and general maintenance. It only took a few months before I realized bare horse hooves and the wet, sodden pastures of my central New Jersey homestead were not compatible. We receive far too much rain in my geographical location and both of my Clydesdales suffered from recurring abscesses. This condition is painful for the horse, causes temporary lameness, and is time-consuming for myself as the caregiver to treat. I didn't have any other options for land for the horses to graze so I had to alter my plan; my horses now wear four shoes each.

A pasture of forage ready for grazing by the first group of livestock. A height of 8" (20 cm) tall forage is desired for grasses, 10–12" (25–30 cm) tall for legumes and forageable brassicas.

Soft green spaces void of rocks or gravel will not help the animal to naturally file down their hooves. Goats and sheep that have access to boulders or rock formations will have the ability to wear down their hooves by way of daily natural activity and will require less human intervention for hoof trimmings. Where soil is absent and there is too much rock, forage will not have the ability to grow and provide adequate calories for animals to graze. Hay will need to be used as a supplement.

Streams and natural waterways may provide a drink for livestock when they are enclosed in pasture spaces. It is worth noting that many municipalities and environmental agencies heavily monitor and regulate livestock access to these brooks and rivers, however. Be sure to contact your city and state before installing any pastures and fencing to discuss ordinances. Environmental protection departments may require buffer zones to keep livestock and water runoff from manure well away from creeks and riverbeds. An open grassy field with a small stream may appear to be the perfect location for a horse or cattle pasture, but city and state zoning officials may not allow it.

One of the questions I am asked most often is with regard to natural plant life and toxicity concerns. It is inevitable that invasive weeds, wildflowers, and herbs will appear in a pasture space. Local state extension offices are usually well equipped with lists of toxic plants by location, and they can help to identify these on your property. While I would not recommend ever turning out horses onto a field full of Queen Anne's lace or lupine, a pasture of lush green grass sprinkled with a few toxic weeds here and there has never been a problem for me. When provided with adequate forage, livestock tend to avoid plants that are harmful to their health as they simply don't smell or taste appealing. When animals are starved of feed and forage though, they are more likely to ingest these poisonous plants as a means to source food. A few unwanted plants within a pasture space can be easily pulled with a well-gloved hand if the animal is not to be trusted.

PASTURE ROTATION

Pasture rotation is the act of dividing a pasture into smaller subsections. Livestock are started in one division and graze for a select number of days before they are moved to the next division. Controlled grazing allows land and forage the time needed to rest and regenerate before animals return.

Timing is everything when it comes to pasture rotation. There is a schedule, an order, and a diligence that comes into play. If stock is kept in one location for too long, the pasture is at risk of overgrazing, which results in slow regeneration and a lack of nutrients for the livestock because their ideal forage already has been consumed. Stock density also is a factor—this is the number of livestock allowed per acre at a given time. More animals per acre means less forage per animal, per acre. Fewer animals per acre will increase the amount of forage they have available to them.

Winter pasture spaces are divided into smaller paddocks for grazing with portable fence lines. This encourages the livestock to evenly graze the entire space, ensuring even growth come spring.

The first step in deciding on the right number of animals for your farm or homestead per acre is to approach your local zoning office. For example, my state currently requires 3 acres (1.2 ha) of land to keep one horse. Each horse thereafter must have an added amount of land equaling 1 acre (0.4 ha). These real estate numbers include both acreage for grazing and for living quarters. When it comes to pasture spaces specifically, livestock population numbers are altered by forage availability, forage types, whether or not hay will be supplemented, whether or not pasture rotation schedules are implemented to speed up vegetation growth, and rainfall—after all, pasture growth requires water. Animal species, breed weight, and pregnancy are other contributing factors. Stock density calculators are available online for helping you determine an ideal balance of animals and forage (see links in the Resources section, page 182).

Once animal populations have been established, it's time to divide the pasture. Many farmers choose the water trough as the point at which pastures are split. This allows for the same trough or bucket to be used throughout the rotation. From there, temporary fencing fans outward, creating a series of smaller paddocks. Division increments should accommodate grazing for two to three days before the animals need to be moved to fresh forage. For two Clydesdales, I allow two to three days on ¼- to ½-acre (0.1 to 0.2 ha) blocks. I have three separate pastures on my homestead; they vary in size from ½ to 2½ acres (0.2 to 1 ha). By subdividing these three fenced in areas into nine smaller plots, I'm able to create a nine-part rotation schedule. If the animals are on each section for three days, that's twenty-seven days of forage before the cycle repeats.

PASTURE ROTATION DIVISION DESIGNS

The following examples show potential pasture division designs for rotational grazing. For ease of explanation, the following pasture is perfectly square. Here are several possibilities for dividing the pasture into four smaller cells. Pastures come in all different shapes and sizes and are not always conducive to dividing up in exact and even increments. Adjust to suit your grazing spaces accordingly.

FAN LAYOUT

Rotational grazing sections use the water trough and gate as the starting point for the divisions. No additional water troughs or entrances are required.

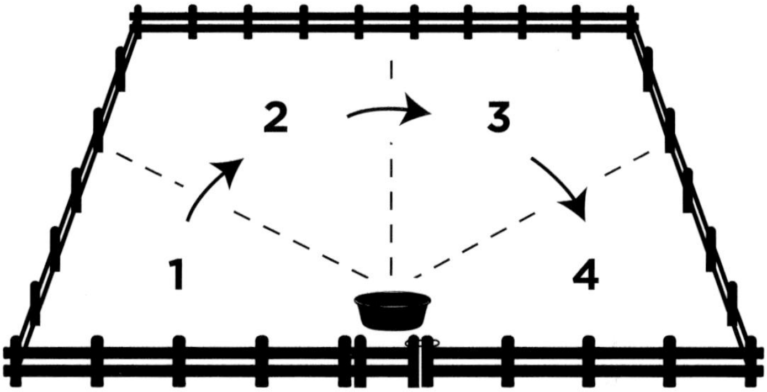

"X" LAYOUT

The pasture is divided into four parts with a large "X." Rotational grazing begins in front of the gate. As the stock moves, so does the water trough. Stock can't exit the pasture without fence removal.

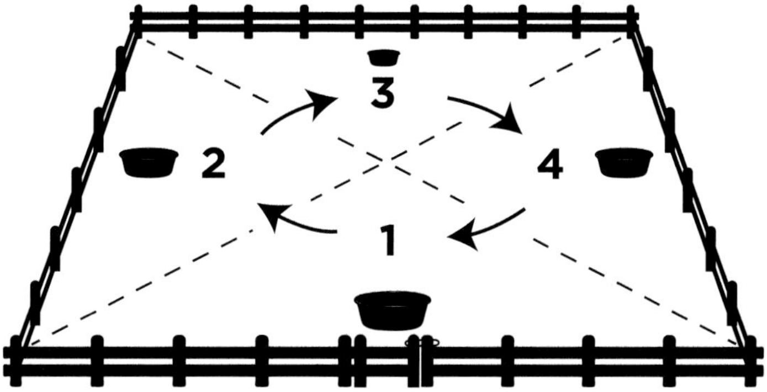

SEQUENTIAL GRAZING LAYOUT

This pasture has been divided into four equal spaces, side by side. As the livestock graze, water troughs and gates will need to migrate as well.

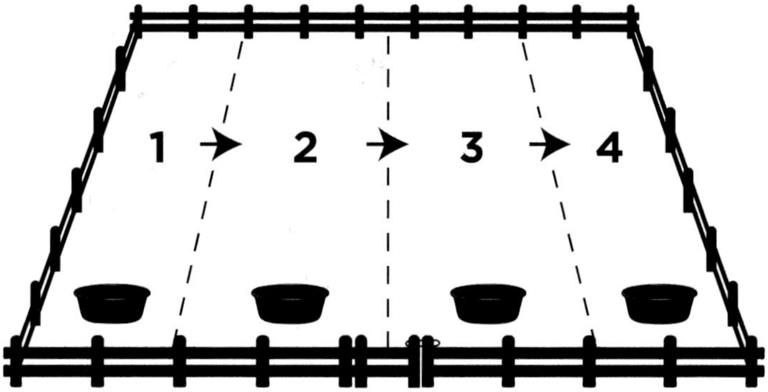

In winter, when forage no longer grows, hay is fed heavily to the livestock. The ducks and geese help disperse hay seeds throughout the pasture and fertilize the cold soil as they go.

Next a pasture rotation schedule can be implemented. One species or more is assigned a paddock for several days then moved onto the next. Tandem or co-species grazing is the act of two or more species grazing side-by-side in the same pasture simultaneously, such as cattle and horses with chickens and goats. A large number of individuals per paddock at any one time means frequent migrations to subsequent paddocks to ensure enough forage is available.

Multispecies rotational grazing requires a little more effort to manage. If more than one animal species is grazing a pasture space separately—the benefits of this practice are discussed on page 116—a follow-the-leader system can keep the cycle organized. Typically, large hoofstock are introduced to a grazing space first when forage reaches a minimum of 8 inches (20 cm) tall for grasses and 10 to 12 inches (25 to 30 cm) tall for legumes. Horses prefer the tender tips of grass blades and would ideally be moved to the next plot after two to three days. If forage falls below the 4-inch (10-cm)-tall desired height for maximum regrowth rates, consider moving the stock a day earlier.

After the horses have left the plot and moved to the next paddock, it's time to introduce smaller hoofstock to the first. Sheep will happily graze remaining grass left by the horses. A second pass of fertilizer is contributed to the soil and the sheep receive the forage they desire. Following the sheep, the grass rests, even though more stock are to be introduced. Goats browse any weeds and undesirable forage discarded by the horses and sheep. They do not graze the grass but continue to fertilize as they browse. After two to three days, the goats are moved to the next paddock where the sheep had just been, and the horses before them.

After the goats, birds are introduced. Geese dutifully weed the pasture of any remaining bits. Chickens break apart manure piles looking for insects that now have had weeks to accumulate. Ducks root around for slugs and snails. After two to three days, the birds are removed and enclosed in the subsequent space. The land rests and regrows, absorbing carbon from the atmosphere, all at an ideal rate because it was not overgrazed. When it reaches ideal height, the horses are reintroduced. If the horses are finished with their rotation before the first paddock is ready, hay can be supplemented.

CLOSED LOOP GRAZING SYSTEMS

More than just soil benefits from pasture rotation methods. The animals themselves graze their ideal forage while constantly being exposed to fresh grazing environments. The eggs of internal parasites are shed in livestock manure. After hatching and several larval stages, the parasites mature enough to become infective. If the host animal is moved before the parasite can corrupt it, there is no host for the parasite to then continue its life cycle. This means the overall population of parasites can decrease.

Parasites are largely species-specific. The stomach of one livestock species can digest and eradicate the parasite that may affect another. Take sheep, goats, and cattle, for example. Cattle naturally consume the larvae of the barber pole worm, which solely affects the stomachs of sheep and goats. Because the parasite is incompatible with cattle, the ingested larvae die. Cattle prefer the taller portions of grass blades. Due to the anatomical formation of the upper lip and smaller head size in sheep, they prefer to graze closer to the ground than cattle where the smaller forage is found.

Barber pole worm larvae climb their way up to the top of the grass blade post-emergence, making themselves readily available for ingestion by their hosts. However, they tend not to climb higher than 4 inches (10 cm) in height, perfect for sheep. One practice of a closed loop grazing system would be to allow sheep to graze a pasture first, taking the tallest portion of grass above 4 inches. The cattle would then follow, eating the lower portions where barber pole larvae reside. But shortening the time for grazing opportunities can also be used for parasite management. If animals are removed from a pasture before the fourth day, parasite hosts have been moved before larvae are able to hatch and then ascend the grass blade for consumption. At my homestead, the sheep follow the horses in the grazing schedule and are given the shorter pass at forage, just as they like. However, they are only allowed in each paddock for two to three days. Barber pole worms do not infect our sheep. Even though the parasite larvae have hatched close to the soil's surface, they have not yet moved up the forage blades.

The benefits of inter-species grazing go beyond cattle and small ruminants. One study explored the effects of co-grazing beef cattle with saddle horses. It was found that shared grazing spaces reduced strongyle egg loads in pastures and ultimately in the shed feces of the saddle horses. Similarly, poultry do not share most parasites with horses or cattle, and they can help reduce parasitic egg counts. Chickens, ducks, and guinea fowl ingest loads of parasitic eggs and larvae when foraging through a pasture. There also is the added benefit of the ingestion of ticks that, otherwise, would affect the livestock.

ANIMAL CONTRIBUTIONS TO MULTISPECIES ROTATIONAL GRAZING SYSTEMS

Each livestock species has its own preferred forage type and contributes different nutrients by way of manure to the soil. When a myriad of animals is allowed to graze in tandem or in a follow-the-leader rotation, the same grazing plot can be used several times to offer each species protein, calories, and energy. This table discusses the preferred forages, placement within the rotation order, and grazing periods recommended for each type of livestock. If grazing livestock side by side, especially larger stock like horses and cattle, shorter grazing periods may need to be implemented to prevent the overgrazing of forages.

WARNING: SICK OR INJURED ANIMALS SHOULD NEVER PARTICIPATE IN ROTATIONAL GRAZING SYSTEMS.

COVER CROPPING FOR ANIMAL FORAGE

My soil had once been part of a cattle farm back in the 1950s. Since then, the land had gone mostly fallow in some areas. In others it had been mowed into a lawn of weeds that were then treated with herbicides. When I finally bought the homestead in 2016, the soil was heavily compacted, void of nutrients, and incredibly difficult to dig. Water was not being absorbed into the soil during rainfall events, and it would run off and pool in pockets of lower ground. It was frustrating to watch all of that water leach away—the soil was always dry and unable to retain any moisture.

When I started keeping horses, I seeded my pasture spaces multiple times per year with Kentucky bluegrass. I wasn't concerned with improving soil structure. My concern was feeding my horses. At the time, I thought these were two separate concepts. For years, I continued to sow grass seed every spring and fall. The horses were given full access to the field and within just a matter of days, a 1-acre (0.4-ha) paddock was mowed to the ground by two Clydesdale horses. The grass never seemed to grow back as quickly as my surrounding lawn spaces, which were lush, green, and mowed weekly. The pasture stayed tightly clipped to the soil's surface, soon dented by the heavy hoof prints of the animals. I remember thinking how strange it seemed that I was giving hay to my horses out in the open field, surrounded by closely cropped, browned grass blades in the midst of summer.

Eventually I came to find that I was managing my pastures all wrong. I wasn't rotationally grazing. I wasn't planting a variety of forage crops to sustain, or even address, soil health—let alone my animals. I was sowing a monoculture of grass that did nothing but prevent soil erosion and absorb a touch of carbon. I had heard of rotational grazing systems on farms, but I couldn't rotate my own pastures until there was proper forage in place for the animals to graze on in the first place. When I learned that cover crops could be both soil fixers and simultaneously animal forage, I had a major lightbulb moment. I found that there were specific cover crop species, such as forage turnip or oil radish, that could be sown into compacted soil. They would loosen the hardened layers as they grew while providing my stock with green tops to feed on. Triticale, a favorite fodder for many livestock species, could offer weed suppression to my pastures, improve topsoil health, and take up and hold soluble nitrogen within the soil. Cover cropping for soil health and providing forage for my animals were not separate concerns; they were one and the same. Furthermore, a variety of cover crops could be mixed and planted to perform a variety of soil-improving tasks at once. A polyculture of cover crops meant a large panel of forage selection for my livestock and greater agro-ecosystem stability. (See page 49 for a list of differing cover crops species and their contributions to soil health.)

Ducks are helpful in jump-starting pasture growth at the end of the winter season. Their droppings fertilize and moisten the soil.

After extensive research, I decided upon a mix of triticale, winter rye, forage turnip, berseem clover, and piper Sudan grass for my grazing pastures. This would be sown in mid-March and in late August. Most of these crops are cold-hardy vegetation and prefer cooler temperatures. Often a blend of grass and legumes are sown together to improve biomass production and mulch thickness, and to create ideal weed suppression. This mix also offers an ideal carbon-to-nitrogen ratio, allowing a gradual release of soluble nitrogen within the soil. An all-grass cover would otherwise tie up this plant-available nitrogen, while an all-legume cover would lead to potential nitrogen leaching and loss.

I decided to sow an organic, heat-loving cover crop mix in early June as a way to provide even more forage diversity to my animals and to my soil. Buckwheat, soybeans, and sorghum-Sudan grass provide a balanced mix of broadleaf annuals, forgeable legumes, and annual grass to generate vast amounts of biomass. Whatever is left of this summer forage after animal grazing is mowed and left to decompose on the soil's surface in late summer. Fall's cover crop blend is sown directly into the chopped and dropped green manure.

When formulating cover crop blends, it is important to understand the plant life cycle of these forages. Warm-season grasses differ greatly from their cool-season counterparts. Cool-season grasses, generally speaking, begin to germinate after sowing in early spring. By June their growth rate begins to slow and seed heads form. Their growth is stunted until autumn when temperatures begin to drop and moisture increases. Very little forage is offered during the summer months from cool-season grasses. This

Axe & Root Homestead is a small farm, measuring only 6 acres (2.5 hectares). Not all this land is dedicated to pasture. Seeding by hand is completely feasible with a seeder attached to a draft horse or an ATV. Here, cover crop seeds are blended and poured into the seeder.

Summer cover crops are carefully selected for their contribution to the soil. They are sown around mid- to late spring here in growing Zone 6b.

is a much longer life cycle than warm-season grass crops. Warm-season grasses begin to grow in mid-late May, grow rapidly throughout the warm summer months and begin to develop seedheads in August or September. While they provide forage throughout the hottest temperatures of the year, they quickly succumb to autumn frost.

A cover crop portfolio that contains both cool- and warm-season grasses will offer viable animal forage throughout the year. Cool-season grasses are heavy feeders and should be grown alongside a legume cover crop to achieve ideal growth. Legumes such as red clover have the ability to fix nitrogen into the soil. Others such as Dutch clover thrive even under intense grazing conditions.

When using cover crops as animal forage, there are many nuances to consider for each animal and each cover crop varietal. For example, wheat is an acceptable forage cover crop for many livestock species. But if sown too early in the fall, such as in late August or early September, it is prone to many diseases that can affect the animal and its overall health. Another example is sorghums, Sudan grasses, and their hybrids; these contain prussic acid (low levels of cyanide). If ingested when fewer than 18 inches (46 cm) in height or after a frost, this can cause irreversible urinary tract disease in horses called cystitis syndrome or cystitis/ataxia. It also has been attributed to abortions and/or birth defects in pregnant mares, and it has a laxative effect as a result of high sugar content. Piper Sudan grass, however, is extremely low in prussic acid and many farmers have grazed their horses on this crop without issue. Furthermore, hay produced from Sudan grasses will not likely cause cystitis/ataxia syndrome as any prussic acid dissipates as the hay is cut, dried, and cured. Because there are so many caveats to grazing animals on sorghum and Sudan grasses, I have included it in the Cover Crops to Avoid column within the Cover Crop Forage for Livestock Species table (page 123).

Species compatibility should always be researched extensively before introducing any cover crop as forage material. Understand the nuances of the crop, and the exact nutrition the animal will be ingesting versus what they require, especially if there are any metabolic conditions. Be sure to discuss these crop possibilities with your veterinarian and animal nutritionist for any/all livestock grazing on the pasture.

The grazing of grass cover crops can begin after the vegetation reaches approximately 8 inches (20 cm) in height. Ideally, legume and brassica forage should minimally measure 10 inches (26 cm). Once the cover crops are grazed to a height of 4 inches (10 cm), it's time to rotate the stock to the next pasture. In one Missouri study, grass was grazed to a height of 4 inches. It was found that a total of thirty days were needed to regrow the forage to a 12-inch (30-cm) height. Comparatively, grass grazed to a 2-inch (5-cm) stubble required forty-five days to regrow to a 12-inch (30-cm) height! This science demonstrates that the overgrazing of livestock on portions of pasture actually slows its ability to regenerate.

As animals graze on cover crops, trampled vegetation is inevitable. These dropped crops are not a loss; they still contribute to soil biology. The following season's cover crop pass can be sown directly into the dropped or crushed green manure. Seeding rates vary based on the crop and are generally measured in pounds per acre. Look to seed manufacturers and package information for precise seeding density.

Cover Crop Forage for Livestock Species

The following table is not intended as a prescription for animal fodder. The information here is provided as a resource for building a cover crop combination that suits the needs of your land and animals.

FORAGEABLE COVER CROPS

Livestock Species	Cover Crop Possibilities	Cover Crops to Avoid
Equine	Berseem clover, annual ryegrass, triticale, winter rye, forage turnip, piper Sudan grass (low in prussic acid)	Buckwheat, brassicas, hairy vetch, sorghum (includes most Sudan grass varieties), millet, lupine
Cattle	Berseem clover, cereal rye, annual ryegrass, wheat, winter rye, oats, triticale, forage turnip, winter peas	Sweet clover, hairy vetch, amaranth, brassicas, flax, sorghum (includes most Sudan grass), millet, corn, lupine
Sheep	Oats, oil radish, forage turnip, field peas, yellow sweet clover, annual ryegrass, winter wheat, triticale	Lupine, hairy vetch, sorghum (includes most Sudan grass), buckwheat
Goat	Ryegrass, winter rye, berseem clover, forage turnip, sericea lespedeza	Lupine, buckwheat, johnsongrass, sorghum, sorghum-Sudan hybrids, Sudan grass, rape, canola, alsike clover, buckwheat
Swine	Alfalfa, ladino white clover, sweet clover, red clover, lespedeza, orchard grass, timothy grass, bromegrass, rape, forage turnip, kale, swede, alsike clover, winter wheat, barley, triticale, cereal rye, crimson clover	Sudan grass, sorghum-Sudan hybrids, flax, lupine
Geese and Ducks	Cow peas, Dutch white clover, winter wheat, cereal rye, barley, corn, buckwheat, black oil sunflower, oats, field peas, ryegrass, triticale, winter rye, alfalfa, red clover, sweet clover, crimson clover, lentils, oats, spring wheat	Hairy vetch, cottonseed, crown vetch, lupine, Sudan grass, tobacco, rape
Chickens	Alfalfa, clover, annual ryegrass, kale, cow peas, forage turnip, mustard greens, oil radish, buckwheat, oat grass, ryegrass, cereal rye, triticale	Rape, lupine

Established cover crops protect the soil from solarization, erosion, and water loss and provide forage for livestock.

MULTISPECIES GRAZING CONSIDERATIONS

Though many parasites may not often infect more than one host species, there are some that are zoonotic. For this reason, good sanitation must be practiced for any shared waterers throughout the rotational grazing process.

Poultry are wonderful for insect and parasite control along with their manure scratching abilities. However, their droppings may carry *Salmonella*. The droppings of infected chickens on livestock feeders, waterers, and any grazing surface can transmit the bacteria *Salmonella enterica* to goats, sheep, cattle, horses, and humans. Care should be taken to ensure poultry manure is completely broken down and removed from water stations upon reentry of livestock.

Some parasitic species such as *Cryptosporidium*, and the bacteria *Campylobacter* (specifically *C. jejuni* and *C. coli*) can, unfortunately, be shared between chickens, goats, sheep, cattle, and humans. Spread through fecal-oral transmission, homesteaders typically see little issue with rotationally grazing these animals unless they share living quarters. Shared soiled bedding, waterers, and animal housing can be difficult to clean once a parasitic infestation occurs. Keep infected animals out of the grazing rotation and quarantined until successfully treated.

It is more likely that a homesteader practicing rotational grazing will suffer from weak fencing than shared parasitic and bacterial infections. When small and large hoofstock share fence lines, be sure to supervise for the first several interactions. Territorial individuals may try to cross the fence line, chase, bite, or jump out altogether. Electrified polywire or netted fencing is a good choice for successfully separating livestock between grazing spaces. Ensure barriers are sturdy, functioning, and strong before introducing any animals.

This pasture has just had all portable fence lines removed from winter turnout in preparation for spring seeding. It's easy to see where the pasture was divided into three sections. The ducks are walking where the horses were several weeks prior, hence longer growth. Above this is the middle section containing overgrazed winter forage. The top area, where the horses currently stand, is the last section on this field's rotation.

MAINTENANCE

The beautiful thing about rotational grazing is that it is relatively maintenance-free. The animals keep the forage trimmed, weeds cleared, and manure spread. The reintroduction of forage seed by way of manure droppings replenishes growth. If Mother Nature cooperates, water is provided. This style of pasture keeping creates a healthy environment for soil and animals and is largely hands-off.

There are several instances where a little intervention is required from the farmer or homesteader. First things first, there's the physical component of actually moving the animals from one paddock to the next. Gates are a hugely helpful addition to any fence line, especially those constructed between adjacent pasture sections. For moving stock to pastures that are completely separated and reside elsewhere on the farm, there's the herding, potential haltering, and leading of animals from point A to B that must be considered.

Seeding may also be required seasonally should the cover crops have thinned due to drought, crop incompatibility, or improper rotation schedules. Moving stock too infrequently results in the trampling, overgrazing, and even death of forage. If pastures are not recovering between full grazing rotations, a new cover crop species may be required.

If chickens are not present or have not scratched through manure piles adequately, a drag can be used. I pull a heavy steel drag behind my ATV or a draft horse when needed. I flip the drag upside down so the teeth point skyward. This allows for spreading manure throughout a green space without the teeth tearing precious soil structure. An even distribution of manure means nutrients are introduced to the soil equally, promoting even forage crop growth.

When birds are not given adequate time per division in a rotational grazing schedule, insects and their larvae may begin to accumulate. Fly predators can be purchased from online retailers for sprinkling through grazing spaces. These creatures are small beneficial wasps that prey on the larvae of horseflies and other manure-breeding pests. They do not bite, sting, or cause harm to humans or livestock, and they are largely nocturnal. Several species make up the trademarked "Fly Predator" family including *Spalangia cameroni*, *Spalangia endius*, *Muscidifurax zaraptor*, *Muscidifurax raptorellus*—all members of the Pteromalidae insect family. They neither directly ingest nor sting their prey. Instead, eggs from females are deposited into larval hosts. The female also takes nourishment from the larvae before moving onto the next.

As with any fenced-in space where animals are allowed to congregate, fencing should be monitored for weakness and breakage. Also ensure sufficient water is provided by way of irrigation if natural rainfall has not occurred. The amount of water depends on the cover crop species. On average, however, 1 inch (2.5 cm) of water per week is adequate for most.

A young goose is happy to graze on weeds and grass left behind by previous livestock groups in this pasture.

CHAPTER six / COMPOST

Where there are animals, there is manure. Where does it all go, and how can it be safely used as growing matter? And how do food scraps fit in?

Compost is a nutrient-rich organic material created by the decomposition of animal waste, yard waste, and other organic items. When water and air are combined with the right amount of green and brown ingredients, the elements decay and are recycled into fresh growing matter. (Earthworms and naturally occurring bacteria heavily assist in this transformation.) Green ingredients are the nitrogen-rich components, such as animal manure, grass clippings, food scraps, or weeds. Dried materials are carbon contributors, and they are referred to as the brown additives. Items such as cardboard, old leaves, sticks or twigs, hay, straw, and paper fall into this category. An ideal compost heap will contain two-thirds carbon or brown materials and one-third nitrogen or green items.

Food scraps that are thrown away and enter landfills account for roughly 30 to 40 percent of landfill contents. As the pressure for more landfill space increases, recycling discarded food items into usable organic matter can offer much-needed relief. Engineered landfills are anaerobic environments where food waste sees very little sunlight and is exposed to little to no oxygen. Food that decays in this anaerobic environment does not break down properly and releases methane gas.

The same is true for animal manure if left to rot in place. With no carbon-based items mixed in, biogenic carbon dioxide is released as the manure decomposes. But if given the proper environment to effectively decompose, the total volume, pathogen and parasite load, and weed seeds greatly reduce. When these nitrogen items are combined with the right ratios of carbon-based items, food waste and animal manure ingredients break down more readily and can be put to use in gardens and orchards.

« Earthworms are heavily present in any active compost heap.

HOT VERSUS COLD COMPOSTING

The most common myths about composting are that it smells like rotting food and manure, and that it attracts flies and wildlife. If properly managed, none of these things should be true. In fact, it more likely is a symptom of a poorly managed compost heap. A pile created with the correct ratios of nitrogen to carbon items will reduce natural odors as items decompose, should not attract flies (especially if hot composting), and should not be a significant attraction for wildlife.

Hot, or thermophilic, composting is a process that consists of three phases. Each phase can be recognized by the temperature range within the heap. The first phase is the mesophilic stage. The pile begins to heat up and reaches moderate temperatures of 50° to 104°F (10° to 40°C). After forty-eight to seventy-two hours, if the pile has enough oxygen and moisture, temperatures within the pile will rise to a higher temperature or thermophilic phase. These temperatures range from 104° to 150°F (40° to 66°C), depending on the ingredients within the pile and the overall size of the heap. This particular stage of decomposition can last from just a few days to several months. The third stage of hot composting is called the curing or maturation period. During this time, temperatures gradually begin to fall. If the heap is turned and oxygen is added, temperatures may rise. This means there is enough organic material remaining within the heap for the microbes to continue decomposition. Once all materials are successfully composted, the pile will no longer generate heat, even after turning.

During the second or thermophilic phase of hot composting, it is extremely difficult for fly larvae, weed seeds, pathogens, and parasitic eggs to survive. A well-maintained hot compost pile can reach peak temperatures of 120°F (49°C) to even 170°F (77°C). Tender weed and hay seeds sporadically begin to die off at roughly 108°F (42°C). It isn't until the heap is heated evenly, using the hot compost method, that grass and weed seeds are destroyed in their entirety. A consistent temperature of 145°F (63°C) over the course of thirty days is required in order to eradicate the most stubborn of weed seeds. These include common groundsel (*Senecio vulgaris*), bird's-eye speedwell (*Veronica persica*), round-leaved or low mallow (*Malva pusilla*), common lambsquarters (*Chenopodium album*), spiny sow thistle (*Sonchus asper*), ladysthumb (*Polygonum persicaria*), wild buckwheat (*Fallopia convolvulus*), field bindweed (*Convolvulus arvensi*) and broadleaf dock (*Rumex obtusifolius*). Any seeds that are not fully hot-composted and cured will likely germinate once the compost is added to the soil within growing spaces.

Finished compost will look like soil. None of the original ingredients should be recognizable.

It is important to remember that exact temperatures and decomposition times are entirely dependent on moisture levels, site location, ingredient physical size, and the overall size of the heap. You can monitor the temperature of your compost regularly with a thermometer. When the temperature begins to decrease, more oxygen is needed to reactivate soil microbes. Give the pile a stir, and the temperature should begin to climb again. Water can be provided by way of rain or hose to keep the pile moist, or a breathable tarp can be used to cover the heap in an effort to retain moisture.

On average, a compost heap measuring 3 feet (1 m) wide and 3 feet (1 m) tall when starting out, should require four stirrings. Over the course of roughly thirty days, from start to finish, the temperatures will rise and fall with each turn of the heap. The volume of the pile should reduce by about half. The compost is ready when its appearance looks like chocolate cake: it should be dark, crumbly, and moist, and none of the ingredients should be recognizable. At this point, the temperatures will remain around 85°F (29°C) or less, even after stirring. Allow the compost to rest and fully cure for two weeks before introducing to the garden.

Cold composting requires much less attention and effort on the part of the homesteader. Both green and brown compost ingredients are heaped and left to decompose on their own. No water is added, and no turning takes place. Materials will take much longer to fully decay—sometimes a year or more depending on the contents. Because the heap does not reach high levels of heat, weed seeds are likely still viable.

Hot or Cold Composting: Which Would I Choose?

Cold composting is certainly easier in terms of hands-on time required. You can dump unwanted food scraps or manure in a pile, add some leaves or other brown matter, and let Mother Nature take control. Eventually compost will remain and can be used just the same as organic matter from the hot composting method. But I much prefer hot composting.

As a homesteader with many growing spaces and thousands of plants to tend to, I require a lot of finished compost to keep my soil structure active, healthy, and full of beneficial microbes. When combined with the output of two Clydesdale horses, there's a lot of manure to be processed every day. For me, the amount of manure my animals generate, and the amount of soil I'm looking to replenish, hot composting is a better fit. With added moisture from natural rainfall, a tarp to keep moisture and heat in, and a few pushes of the heap with my ATV's front plow blade, I can successfully have compost from start to finish in just a few short months. With a fast turnover rate, there's also less real estate needed for various cold compost heaps.

Duck compost is one of the few fresh manures that can be directly applied to the garden without composting first. Even so, we compost all manures and soiled beddings at the homestead.

Waste from horses, cattle, goats, sheep, and other farm animals is generally safe for composting. However, dog and cat waste is best avoided, especially if the compost is to be used in growing spaces designated for food production.

What Not to Compost

By nature, just about every organic material will eventually decay if left to rot. There are some items, however, that are best not added to the compost heap, and others that are a judgment call:

- Dog and cat waste is loaded with harmful pathogens and parasites, such as roundworms and the organism that causes toxoplasmosis.
- Meat, fish, and dairy items naturally attract flies and a large variety of wildlife for a feast as they decay. They also contain bacteria including *E. coli* and *Salmonella*. It can be difficult to fully remove these bacteria from a heap and, if used on crop growing spaces, the finished compost could transfer bacteria to edible plants. Unless there is a guarantee that the entire compost pile has reached temperatures ranging from 141° to 145°F (61° to 63°C) which would terminate those harmful elements, then dairy, meat, fish scraps, and animal waste are best left out of the pile.
- Citrus peels are a welcome addition to the compost heap, however, the fruit itself contains high levels of acid. Citric acid can kill helpful microbes and earthworms within the pile, slowing activity.
- Raw eggs are a green ingredient that would add nitrogen to a compost heap, but they produce a strong, off-putting odor to humans while rotting. The same scent is very appetizing to wildlife, however, and can attract unwanted visitors to an open pile.
- Plants that have been treated with herbicides or are diseased require very high hot compost temperatures in order to successfully die off. Without a guarantee that the entire pile has been equally heated, these ingredients have a strong likelihood of infecting an entire compost pile. That compost would spread those elements back into your growing spaces.

Compost Additives

Compost can be customized to best suit the needs of your crops and gardens. Start with a base recipe of 1 part nitrogen or green items to 2 parts carbon or brown materials. Add the following for the boost you're looking for.

FOR A POTASSIUM (POTASH) BOOST

Potassium helps with overall plant growth and resistance to drought and disease. Also stimulates the stomata opening and closing in plants, regulating the uptake of carbon dioxide through photosynthesis.

Add: Citrus peels, wood ash, banana peels

FOR AN INCREASE IN CALCIUM

Calcium promotes healthy plant tissue and membrane development. Also helps root enzyme development within the soil.

Add: Eggshells, wood ash, oyster shell, gypsum, bonemeal

IRON CONTENT BOOSTERS

Iron helps the plant with all major functions, including photosynthesis, chlorophyll production, nitrogen fixing, and enzyme production.

Add: Blood meal, animal manure, lentils, chickpeas

FOR PHOSPHOROUS-RICH COMPOST

Phosphorous is important for root growth, flower, and fruit development.

Add: Bonemeal, animal manure, clay, wood ash

TO INCREASE MICROBIAL ACTIVITY

Microbes are what speed up the decomposition process of decaying materials within the heap. A healthy amount of microbes will transform the heap from ingredients to finished compost more quickly.

Add: Sourdough starter discard, wine, beer, kombucha SCOBY

In mid-spring, the compost heaps will be depleted after organic matter has been removed and spread in growing spaces.

CREATING A SYSTEM

There are several factors to be taken into consideration before installing a compost system. Site location, water availability, rotation ability, and retrieval access all heavily influence the finished product. Keep in mind, though, that the process of composting is a natural transformation. Heavily constructed or sophisticated systems certainly can make the process faster or the act of rotating and turning easier, but they are not required. A pile of compost will decompose directly on the soil's surface just like a heap contained within well-built walls.

Site Selection

A flat, level surface with good drainage is required to avoid water and nutrient runoff. Keep the heap at least 40 feet (12 m) away from natural waterways to prevent accidental infiltration of compost ingredients. The system should be easily accessible but well away from the home. If animal manures are being added, keep the heap away from animal living and grazing spaces as well to avoid manure and active parasitic larvae runoff in rainy weather events. Exposure to sunlight is helpful in heating the heap, as is accessibility to a water source—especially if you reside in a dry climate. Locations with plentiful rainfall may not need to moisten the compost heap, but for those with little to no rain, irrigation will be necessary to keep the pile wet. After the compost is finished, close proximity to growing spaces is helpful to reduce effort and transportation.

Consider Sections

A multi-bin compost system is helpful in organizing a large quantity of compostable material. One slot or bin can be filled at a time. When the pile has reached its ideal height, the second section is added to, and so on. This allows for multiple heaps to be actively composting while others are just beginning or are in progress. This is particularly advantageous when cold-composting methods are used and heaps can take long periods of time to fully decompose and cure.

Access

Wooden slats or sideboards are aesthetically pleasing and can help to keep a compost pile well contained. But how will the finished product be retrieved? Is there a door system? Do individual side boards need to be removed? Can a wagon pull up alongside the system? If not turning the compost in a hot compost regimen, then remember the bottom of a cold compost heap will be harvested first. Access to the bottom of the pile will be required in order to use the finished organic matter.

Air Circulation

Cold composting does not require turning as items are left to decompose in place. Hot composting offers a finished product much faster, but requires regular turning or stirring to stimulate microbial activity. If hot processing your compost, how will the heap be turned? From the side with a shovel or front loader? This can become heavy and laborious work.

A method for ensuring that the centermost compost ingredients receive plenty of oxygen is to use an air tube. Air tubes are simply homemade ventilation pipes that allow air access. A 4- to 6-inch (10- to 15-cm) diameter tube or PVC pipe can purchased at most hardware stores. Drill holes into the sides of the pipe at various locations, each hole measuring ½ inch (1 cm) in diameter. The length of the entire pipe should be inserted into the center of the compost pile and extend outward from the pile at least 6 inches (15 cm). This allows plenty of air to enter the pipe and perforate through the sides into the heap, speeding up the decomposition transformation.

A two-tier compost system. Manure and compostable material are piled into one section at a time until the area is filled. While one section decomposes, the other is added to.

USING COMPOST ON THE HOMESTEAD

Compost is not a fertilizer. While it does contain beneficial macro- and micro-nutrients for plants, it is not plant food. Instead, compost should be thought of as a soil amendment or conditioner. Compost is organic matter that is meant to contribute to soil structure, fertility, and stimulate microorganism activity within the rhizosphere (see page 128 for more information on soil). When these mycorrhizae and microbes are present, they help stimulate the uptake of water and nutrients by the plant's root systems. Compost is a facilitator of plant health and function, not a food source. Studies have shown that surface-applied compost applications in growing spaces improve water absorption, overall tilth, and organic material. Nitrogen is not readily available for plants to absorb, and manure-based finished composts contain a moderate amount at best (roughly 2 to 4 percent nitrogen). Vegetative waste-based composts contain even less nitrogen at approximately 1 to 3 percent on average.

The nutritional value of compost in terms of nitrogen, phosphorous, and potassium is different for every heap, as finished nutrient levels are directly related to what materials were used to create the compost in the first place. Generally speaking, if the heap was well balanced in terms of nitrogen to carbon ingredients, it will hold a fertilizer content of approximately 1-1-1 (one part nitrogen, one part phosphorous, one part potassium) with an average pH of 7.

Much of the nitrogen in compost is not readily accessible by plant roots, as it is tied up within organic matter. Soil microbes mineralize the organic nitrogen over time and convert it to ammonium and nitrate, which plants can readily absorb. Therefore, compost should be thought

Compost Troubleshooting

Composting is a naturally occurring process that takes time and a balance of ingredients. If not enough nitrogen-based items, carbon-based ingredients, water, or air are provided, an imbalance will make itself known through odor or lack of decay. A compost heap that smells rancid like vinegar or rotting eggs is a symptom of a lack of oxygen or too much nitrogen. Check to see if the pile is too soggy; if so, turn to aerate the heap. Add sawdust, wood chips, leaves, or other carbon-based materials to achieve a better balance.

When the compost pile displays a lack of heat, especially when conducting the hot compost approach, the pile may be too small or too dry. To reach the desired temperatures, a heap must contain enough organic material to measure at least 3 feet tall (1 m) and wide. If this is not the cause, try adding more moisture to the pile. A pile that is wet but doesn't generate heat may emit a sweet scent. This could be from a lack of nitrogen-based materials. Try adding more food scraps, grass clippings, or manure.

If the compost pile begins to attract wildlife, food scraps are not sufficiently covered. Be sure to coat any food waste with brown, carbon materials (e.g., leaves, straw or hay, and sticks). Meat and dairy products are typically the main attractants for raccoons, skunks, and opossums.

Before introducing compost to any growing spaces, it is carefully examined to ensure the contents are fully decomposed.

of as a slow-release source of nitrogen for crops. The same is true for phosphorous. After one year, only 35 percent of the phosphorous within finished compost is readily available to plants. It will generally take a time period of three to five years for 100 percent of the phosphorous in compost to be accessible. Potassium, however, is not generally tied up within organic matter. Therefore, compost is considered a readily available source of this major nutrient for crops.

Before fresh compost is applied to gardens and crop growing spaces, it's essential to ensure the compost is actually finished decomposing. If unfinished compost material is applied and it is loaded with carbon-based items, the compost could absorb nutrients away from surrounding soils in an effort to fulfill its own demand to finish the transformation process. Alternatively, if too many nitrogen-based ingredients are contained within the unfinished compost, there is a likelihood that the high amounts of ammonium contained within those feedstocks could burn or harm plants. Allowing several weeks for compost to cure after it is deemed finished is an insurance policy that all of the materials within the organic matter have fully decomposed.

Growing Spaces

The most common method for applying finished compost is to spread an even layer directly on top of garden beds, on-ground crop soil, and around fruiting trees. A surface application of approximately 1 inch (2.5 cm) in height can help improve water absorption and retention and soil structure; it also can introduce beneficial contributors, such as earthworms and microorganisms. I personally do not rake or till my compost amendment into the topsoil as this disrupts active soil structure. As the compost is moistened, nutrients from the organic matter will leach downward into the topsoil and any existing root systems.

When planting trees or shrubs, transplanting seedlings, or installing flowering bulbs, a bit of compost can be sprinkled directly into the hole before planting. This action will jumpstart microbial activity surrounding the newly introduced roots and has been reported to increase bloom times, harvest yields, and overall health of the plant. Because compost is considered a soil enhancement and not necessarily a fertilizer, any macro- or micro-nutrients within the compost are slowly leached into the rhizosphere assisting nutrient uptake and water absorption on behalf of the newly planted crop.

One of my favorite ways to employ an excess of compost is to sow my pumpkin, gourd, and melon seeds directly in or around a finished compost heap. These plants are heavy feeders, meaning that the require consistent levels of nitrogen, phosphorous, and potassium over a long period of time. Compost can provide this even, slow release of nutrients and greatly facilitates absorption.

Compost Tea

Plants absorb nutrients in liquid form faster than when in mineralized or solid form. Compost tea is a way to capitalize on this natural feeding process. Compost is combined with water in a large bucket or container. The contents are allowed to steep for twenty-four hours, undisturbed, to fully infuse the water with the contents of the compost. The tea does not require straining if brewed in loose-leaf form as all of the components are beneficial to the garden. Unless a foliar sprayer is being used to administer the tea and contains a nozzle with small holes, the tea can be fed in its full, freshly brewed state.

Homemade compost tea can be made using a large 5-gallon (19-L) bucket, a cheesecloth or unbleached coffee filter, and tap water. If living within city limits, tap water that is chlorinated will require a curing period of twenty-four hours before use. Research has shown that low levels of chlorine in tap water can kill beneficial microorganisms within the compost. A bubbler can be helpful when brewing tea and offers a superior product to tea brewed without aeration, though it is not required. Introducing oxygen into the tea stimulates beneficial organism growth and function and can increase the population of microbes beyond the compost feedstock initially used to make the tea. It also is said to reduce pathogen numbers as these harmful organisms thrive in anaerobic environments. Compost tea created using a bubbler is referred to as Actively Aerated Compost Tea (AACT). If brewing compost tea without an aeration system, it is of the utmost importance to ensure that the compost being used for the brew was not derived from pathogen-carrying ingredients.

When applied to the garden, a weak fertilizer is delivered. But more importantly, the beneficial microbes including fungi, nematodes, bacteria, and protozoa within the compost are added to the growing space. When a foliar sprayer is used to drench the plant's leaves and stems, their tissues are armed, giving them a layer of protection against some pests and disease. If drenched onto the soil, root growth and soil structure is stimulated. Regular compost tea applications have been said to loosen clay or compacted soils, to help sandy soils retain moisture, and to replace soil organisms that had been previously lost as an overuse of fertilizer and herbicides.

Tea should be applied no more than once every fourteen days to avoid excess nutrient runoff. Too much compost can leach away into surrounding spaces resulting in a buildup of nutrients or natural waterway contamination. This buildup promotes unwanted growth in the form of algae.

Basic Compost Tea Recipe

MATERIALS

- 1 cup (122 g) finished compost
- 1 cup (100 g) garden soil
- Cheesecloth or unbleached coffee filter
- String
- 5-gallon (19-L) bucket
- Tap water (if water is chlorinated, allow to sit for twenty-four hours)
- Aerator (optional)

DIRECTIONS

1. Combine the compost and garden soil on a cheesecloth or unbleached coffee filter. Secure tightly by tying with a string. Submerge the tea bag in the bucket filled with tap water. A water temperature of 68° to 72°F (20° to 22°C) is ideal. Add any desired additives to the water. If using an aerator, place in the bucket and turn it on. Brew the tea for twenty-four hours.
2. Dilute the finished tea to 3 parts tea, 1 part water before applying to growing spaces. Early morning or cloudy day applications are recommended to avoid sunburned foliage. Try to use the tea within two hours of brewing to deliver optimal benefits.

« Homemade compost tea has been diluted and is being applied to individual crops within the hoop house by way of a watering can.

OPTIONAL COMPOST TEA RECIPE ADDITIVES

Liquid compost, in the form of compost tea, delivers soil boosting organisms and trace minerals to plants in a readily accessible form. Tea can be customized to meet the needs of your crops with the following additives.

Worm castings: Use in place of compost altogether, or use ½ cup compost, ½ cup worm castings in the recipe provided. Worm castings contain more humus than compost or garden soil, allowing for higher levels of water and nutrient absorption. Also contains active soil microbes and low levels of iron.

Unsulphured blackstrap molasses: 2 tablespoons (30 ml) offers a food source for beneficial bacteria, stimulating their population growth. Also fortifies the tea with a source of iron that won't burn plants.

Liquid fish emulsion: 1 teaspoon of liquid fish provides a readily accessible source of nitrogen, phosphorous, and potassium. Also stimulates healthy soil structure.

Liquid kelp: 1 teaspoon of this renewable resource added after brew is complete stimulates chlorophyll production, overall plant growth, photosynthesis, and strong root development. Also assists with nutrient absorption and contains trace amounts of nitrogen, phosphorous, and potassium for plants.

Epsom salt: 1 teaspoon is a source of magnesium for plants that assists with flowering and fruit production.

CHAPTER *seven*

THE SUSTAINABLE ORCHARD

In my backyard, there are two towering black locust trees rivaled in height by a neighboring silver maple and a white oak. Within close proximity reside an old pear tree of unknown variety, a crab apple, and the remnants of an old Winesap apple tree that fell during a hurricane. Below the fruiting trees are two clusters of Russian olive shrubs, a lilac bush, and a buckeye. The soil in this area is soft compared to the hard clay that surrounds the perimeter. The trees are healthy, home to birds and insects, and—despite their age which a local arborist estimated to be around a hundred years old or more—still produce fruit. It wasn't until I became familiar with permaculture orchard practices that I realized this entire area was a functioning micro-forest.

Wooded landscapes within nature consist of multiple layers of growth with diverse vegetation, or polycultures. The tallest canopy creates an overstory that provides dappled shade and shelter to all of the inhabitants below. Beneath the overstory resides the mid-story, a tier comprised of trees smaller in height. The understory is home to small trees and shrubs, which remain low to the forest floor. Groundcovers and vining plants spread outwards along the soil's surface, weaving between herbs, flowers, and companion plants. A grower can mimic this natural system of vegetation by creating a food forest garden (find more information in chapter 3). Another place to replicate this growing pattern is within the homestead orchard.

« Even dwarf trees, such as this cherry, provide abundant harvests year after year. The physical fruit is full size; it is only the yield that is smaller than semi-dwarf and standard counterparts.

CREATING AN OVERSTORY

When most folks hear about home orchards, they picture vast rows of apple trees. However, companion planting overstory trees with fruiting tree species has its advantages. Tall hardwood trees can provide a harvestable crop for the homestead. A few examples of these trees include maple, oak, walnut, and chestnut. Their wood can be used for lumber while acorns, sap, or nuts can be harvested for use. In addition, these varieties grow quickly, absorbing massive amounts of carbon dioxide from the atmosphere as they undergo photosynthesis, storing it within their fibrous tissue. Their expansive root systems pull carbon downward where it is sequestered within the soil.

Massive trees provide homes to birds, insects, and wildlife. Their large branching structures create windbreaks and provide shade, resulting in cooler temperatures below to residential housing and animals. They also create natural support trellises for vining crops. Harsh winds and sunlight can create challenging growing conditions for some saplings. When planted alongside hardwood companions, these trees act as nurses to young shoots, giving them a protected growing environment to establish their roots.

The Moringa tree is one of the most beneficial overstory trees in warm weather climates. Every single part of the tree is edible, from its leaves to its roots, containing massive amounts of vitamin

Mid- and overstory tree canopies provide the farm with shade, support for other trees and crops, and food. Left to right: Russian olive tree, oak tree, apple tree, silver maple tree, crab apple tree, elm tree, black locust tree.

C, antioxidants, and minerals. In addition, this species is drought tolerant once the tree is well established. This fast-growing tree acts as a beneficial companion to any neighboring plant life growing around it. As a member of the legume family, it fixes nitrogen into the soil, improving fertility and structure. Just as with other tree species, the expansive roots assist in erosion control while holding the soil in place. But studies state that the Moringa absorbs carbon at a rate twenty times higher than that of most general vegetation, storing absorbed gases in both its tissues and sequestering it within the soil. Unfortunately, the weakness of "The Miracle Tree" (the nickname given to Moringas around the world) is its intolerance for cold temperatures. While this tree may not work for northern growing climates, there are many cold-loving overstory options available.

Which Overstory Trees to Grow

There are environmental considerations to account for when deciding which trees to plant within the homestead orchard. Your growing zone will determine which varieties are best suited to the temperatures and available rainfall of your climate. Also worth noting is the available space in your homestead orchard, yard, or growing space to accommodate a large overstory tree.

Black Locust

Just as the Moringa is known as a fertilizer tree, so is the black locust. This species has the ability to pull nitrogen from the atmosphere and fix it within the soil, creating a readily available source for neighboring plants. While some may consider this to be an invasive species, this fertilizer tree is full of function and contribution to the homestead. If seed spread is properly managed, the black locust can serve so many other purposes. When coppiced, the wood itself is a long-burning wood and generates much heat. This provides a source of warmth to homes and structures relying on woodstoves throughout the winter. Black locust is naturally rot resistant and contends with oak for its strength and overall durability, making it an excellent choice for outdoor construction and woodworking. The seed pods, leaves, and bark are toxic to humans, cattle, sheep, and horses; the flowers are considered a delicacy in the restaurant food industry—and also are a favorite of honey bees and pollinators. Goats, however, are happy to browse all portions of the tree and keep spread under control.

Black locust trees are relatively drought tolerant; they can survive a lack of water, but their nitrogen-fixing abilities slow. One study found that an entire forest that was decimated by the logging industry, sprouted many black locust trees and they appeared to thrive despite the harsh disturbance. As a result, the soil was enriched from the nitrogen, and the soil was held tightly in place by the roots, preventing further erosion. Soon other tree species were able to establish themselves and regeneration of the forest plot was attributed to the black locust tree. With its ability to provide large amounts of biomass, tolerate soils within a wide range of pH levels, thrive in various growing zones, and tolerate juglone (a toxic compound released by walnut trees), black locusts can serve as a productive buffer or barrier between walnuts and other crop species.

Chestnut

The American chestnut was once highly prevalent throughout the eastern portion of the United States. These massive trees were among the tallest, fastest-growing, and strongest trees to be found in deciduous forests, and they provided lumber that was rot resistant and straight-grained, perfect for exterior construction and beautiful enough for furniture and indoor woodwork. The nuts of the tree served as a major food source for humans, livestock, and wildlife, providing a 20 million pound (907 metric ton) crop, a massive source of income for chestnut farmers and for those who fattened their livestock on the meat of the nuts before processing. Sadly, the American chestnut tree fell victim to a deadly blight introduced from Asia around the year 1900. While the root structure of the tree has not died off completely and new saplings do emerge, they eventually all succumb to the disease and die. Functionally speaking, this ideal permaculture tree species is now considered extinct.

The Chinese chestnut tree is a modern-day substitute to the historic multipurpose tree species. Though it only reaches roughly 40 feet (12.2 m) in height (as opposed to the American chestnut which averaged approximately 100 feet [30 m] tall), a crop is ready for harvest just four years after planting. The trees are very productive and harvest size continues to grow with each season, providing forage for wildlife and livestock, and for human consumption. Though the tree prefers well drained soil, the Chinese chestnut can grow well in sand, clay, and nutritionally lacking locations. The spiky pods contain the nuts. They are painful to the touch, but they can be used as fire kindling when dried. Chestnut trees require, on average, 300 to 500 chill hours in order to successfully create fruit each season (more on chill hours on page 151).

Oak

Wildlife and insects favor oak trees for their ability to provide food and sanctuary. Caterpillars feed on the small blossoms and leaves while squirrels, deer, and other wild foragers scavenge for acorns. Hollow cavities within the trunk are a favorite home preferred by honey bee swarms, owls, and other nesting birds. The vast branching systems provide shade and much-needed relief from the hot sun for livestock or any wildlife animals who choose to take shelter beneath. A relatively rot-resistant wood, even members of the fungi kingdom prefer to sprout mushrooms on the sides of oak trunks and on fallen logs.

Humans can benefit from oak trees by feeding acorns to pigs (more about the controversial practice of pannage in chapter 4). Fallen nuts can be gathered and used for acorn coffee, flour, and ground acorn meal after tannins are successfully removed. Oak timber, specifically wood from the white oak tree, is considered to be one of the heaviest woods, and is rot resistant. The wood from pruned or fallen branches can be cut and used for firewood, mushroom farming, or wood chips. Oak wood is an excellent choice for furniture making, construction projects, and fence posts. The tannins from acorns have been used in animal hide tanning, and the bark of the tree is said to host medicinal properties for a wide variety of ailments. Fallen leaves are an excellent mulch in home gardens, returning nutrients and organic matter to the soil while helping to retain moisture.

Through photosynthesis, oak trees improve air quality by intaking carbon dioxide. Oak trees have the potential to live up to 400 years and can absorb roughly 92 pounds (42 kg) of carbon dioxide annually. That's a possible 36,800 pounds (16,692 kg) of CO_2 removed from the atmosphere and sequestered from just one oak tree. Wineberries, blackberries, ginger, rhubarb, and horseradish are great companions beneath an oak.

Maple

Perhaps the most well-known of the maple trees is the sugar maple, as its sap is coveted for maple syrup. But any maple variety has the ability to be tapped for sap—it's the sugar content within the sap that varies. And these hardwood trees offer more than a wintertime food source. Their blossoms appear before leafset, and are a favorite of honey bees and native pollinators. The leaves and their stems are consumed by deer, moose, hare, squirrels, chipmunks, porcupines, and various insects, and also create an excellent mulch for landscapes and growing spaces when dried. Note that dried leaves are toxic to horses and some livestock, so shouldn't be allowed to drop in pasture spaces. Seed pods contribute to the diet of squirrels, chipmunks, and various bird species. The wood of the tree is a popular choice for its durability and contains a straight grain that is ideal for butcher block countertops, flooring, furniture, and cabinetry. The bark is said to be medicinal while containing tannins that are slightly astringent.

The sturdy trunk and canopy of maples provide vast amounts of shade during seasons of harsh sunlight, along with protection from harsh winds, while offering permanent shelter to a myriad of animals. Sugar maples can live up to 400 years, red maples roughly 300, and silver maples approximately 100 years. Silver maples, one of the most effective absorbers of carbon dioxide, can take in 25,000 pounds (11,340 kg) of CO_2 by the time it reaches age fifty-five years old. That's a massive amount of carbon removed from the atmosphere, per tree, in one life-span.

Walnut

Walnut trees are not necessarily good companion trees—especially for nightshade family crops including tomatoes, peppers, eggplant, potatoes, and also cabbage. But within their own polyculture guilds, and when surrounded by juglone-tolerant plant life, walnuts can provide large canopies resulting in filtered shade, edible nuts, and hard wood for lumber. Juglone is the toxin produced by walnut tree varieties, butternut, hickory trees, and pecans; it inhibits the growth of any nearby vegetation. Black walnut contains the highest concentration, found in the leaves, branches, and fruit. Through root exudates juglone can persist within the soil for months, long after a tree and its roots or stumps are removed. A good rule of thumb is to plan for the toxin to be present for up to a 50-foot (15-m) radius from the trunk of the tree.

Peaches, cherries, nectarine, and plum trees seem to not be bothered by the effects of walnut tree species. However, it is not recommended to plant this overstory tree near apples or pears, as they can be juglone-intolerant. Though they may have limitations with regard to food forest incorporation, walnut trees are extremely versatile. The actual walnuts themselves are filled with minerals and provide a source of food (and income) to humans. Livestock and insects take refuge within the tree and feed on the meat of the nuts. Ground hulls can be use in natural dye baths and ground nuts are used in grit paper and exfoliant production. As an extremely hard wood, walnut is the lumber of choice for cabinetmaking, paneling, and furniture. It is worth noting that while cultivars vary, on average a walnut tree requires 8 to 100 chill hours in order to set fruit.

Because these trees grow at a fast rate with a life expectancy of around 200 years, walnuts are capable of absorbing large quantities of carbon dioxide over the course of their lifetime. They are relatively drought tolerant and have sustainable root structures for flood survival. Calendula, zinnias, carrots, squashes, corn, melon, and beans are excellent companions for walnut tree growing spaces.

Tapping Trees for Syrup: Varieties

Did you know that you can tap many different varieties of trees for sap? This sap can be boiled down into delicious homemade syrup. Maple and birch trees are most common, but there are many other choices for syrup making. The sap from each tree holds different amounts of sugar and water; this means more sap may be required to make syrup. The taste also will vary. Sap flows at different temperatures for different trees as well.

Maple (sugar, silver, black, red, Norway, big leaf)

- 40 parts of sap yields 1 part of finished syrup
- Tap when daytime temperatures are above 32°F (0°C) and nighttime temperatures are below

Birch (European white, paper, yellow, black, gray, river)

- 110 parts of sap yields 1 part of finished syrup
- Tap when daytime temperatures are 40° to 50°F (4.4° to 10°C)

Box Elder

- 60 parts of sap yields 1 part of finished syrup
- Tap when daytime temperatures are above 32°F (0°C) and nighttime temperatures are below

Black and English Walnut

- 60 parts of sap yields 1 part of finished syrup
- Tap when daytime temperatures are above 32°F (0°C) and nighttime temperatures are below

Butternut

- 60 parts of sap yields 1 part of finished syrup
- Tap when daytime temperatures are above 32°F (0°C) and nighttime temperatures are below

Sycamore

- 40 parts of sap yields 1 part of finished syrup
- Tap when daytime temperatures are above 32°F (0°C) and nighttime temperatures are below

Palm

- 88 parts of sap yields 11 parts finished syrup
- Can be tapped year-round

Gorosoe

- 40 parts of sap yields 1 part of finished syrup
- Tap when daytime temperatures are above 32°F (0°C) and nighttime temperatures are below

Other edible tappable tree varieties include linden/basswood, ironwood, alder, and more. Pine trees can be tapped, however, their sap is very resinous and used for resin and turpentine.

At Axe & Root Homestead, all maples on the property are tapped for sap to be boiled into syrup. This includes sugar maples, silver maples, and Norway maples.

Tapping Trees for Syrup: How To

Regardless of the variety of tree you are tapping, the process is the same. Be sure to always tap trees that measure 10 inches (26 cm) in diameter or more so as not to damage the heartwood. A tree measuring this size can withstand one tap. A tree measuring 20 inches (51 cm) can handle two. Finally, a tree measuring larger than 25 inches (63 cm) in diameter may have three taps. Never install more than three taps per tree. When installing multiple taps, always place them at a minimum of 6 to 8 inches (15 to 20 cm) apart from one another.

Materials

- Power drill
- 5⁄16-inch (0.8 cm) drill bit
- Spiles
- Hammer
- Bucket hooks (if hanging buckets)
- Hoses (for ground buckets)
- Buckets with lids
- Harvesting storage buckets
- Large pot
- Thermometer

Directions

1. To tap the tree: Locate the south-to-southwest side of the tree. Drill a hole at a slight upward angle, approximately 1 to 1½ inches (3.5 to 3.5 cm) deep. If using hoselines, be sure your hoses can reach from the location of the tap to the bucket.
2. Insert a small twig into the hole and gently clear out any shavings. Insert the spile and gentle hammer into place.
3. If using hooks, add them now and then hang your bucket. If using hoses, add them now and install your bucket at the base of the tree. Bucket lids are recommended to keep out rain.
4. Check the buckets regularly and collect the sap. Store the sap outdoors to avoid fermenting. When you have accumulated at least 5 gallons (19 L), you are ready to boil.
5. To boil the sap: Build an outdoor fire and create a grate for the pot to rest on. We use three industrial steel pipes over the span of our fire pit. Place the pot on the fire. Pour the sap into the pot.
6. Allow the sap to boil. As it steams, the water will evaporate. It is recommended to do this outside as this steam contains a sticky residue that will coat your walls. When the sap has reduced by at least 75 percent, bring it indoors to finish over the stove.
7. Allow the sap to boil on the stove top until it reaches 218°F (103.3°C). This is the point when the sap turns into syrup. If at any point during the boiling process the sap begins to bubble over the top of the pot, a small pat of butter may be added to reduce the foam.
8. Filter the hot syrup through a paper cone filter, cheesecloth, or fine-mesh sieve into clean canning jars and allow to cool completely.

THE MID-STORY: FRUITING TREES

Imagine planting a seed once and, with proper care and the right growing conditions, receiving a harvest annually for decades. This is the power of perennial crops such as fruiting trees. Incorporating perennial crops into the homestead increases organic matter within the soil and improves overall structure, absorbs and sequesters carbon dioxide, provides harvestable yields that are available annually and usable wood or timber (more on the benefits of incorporating perennials with regard to soil health in chapter 2). Birds, insects, and wildlife seek shelter in these permanent trees while honey bees and native pollinators seek their blossoms as a source of pollen.

In the right location, a fruit or nut tree can produce a crop abundantly. In the wrong location with too little sun or inadequate water, lush foliage may appear unaccompanied by a harvest. Typically, fruiting trees prefer full sun (at least six hours or more per day), plenty of irrigation, good drainage to prevent rotting of the roots, and adequate space to create a large, expanded canopy. All of these considerations should be taken into account when planting. When interplanted with overstory trees, the mid-story should still receive plenty of dappled sunlight throughout the day.

Planting depth and hole size are determined by the exact tree species and size grouping (see the size and yield section below). Trees are best installed in late fall or early spring when the ground is not frozen but the tree is dormant. This allows the energy of the plant to focus on the roots, which should establish before temperatures warm. By the time buds begin to swell and leaves are setting, a sound root structure will have been created. Generally speaking, most fruiting trees thrive in a loose soil with a pH of 6.3–6.6.

Our historic homestead is home to many massive, century-old fruit trees. To keep them in production, they are pruned and maintained by a local arborist.

Which Fruiting Trees to Grow

All fruit and nut trees have specific growing requirements when it comes to temperature, pollination, and spacing. An apple tree of any variety may grow in both warm and cool climates, but it will not produce fruit equally. Fruit and nut production are dependent on the following factors.

Chill Hours

Many fruiting tree species require chill hours in order to successfully set fruit each spring season. One chill hour is any one hour a tree experiences when temperatures fall below 45°F (7°C) but remain above 35°F (2°C). Some trees require more chill hours than others. Apples are a tree species that generally heavily rely on chill hours for adequate fruit production. If sufficient chill hours are not reached, trees will leaf out later and have a prolonged blossom period. This longer bloom time will open the tree up to disease.

For growing regions with a minimal number of chill hours, low-chill varieties may be available. Most persimmons, berries, olives, and pomegranates prefer fewer chill hours, while pears, apples, plums, cherries, and peaches tend to require more.

Perennial harvests increase with each passing year.

Apple trees provide the homestead with fruit, wood from fallen or pruned timber for firewood and wood-working projects, and also homemade apple cider vinegar and other goods.

Peach and other orchard fruit trees produce blossoms that, if successfully pollinated, will ultimately create fruit. While some trees are self-pollinating, all trees will produce a higher yield with a pollinating partner.

Peaches and other fruit are sorted for preservation. Any bruised, diseased, or blemished fruit will be saved for fresh eating. Only flawless items are packed for canning and storage.

Pollinator Groups

Some fruit and nut trees are able to pollinate themselves in order to create fruit. This is what the term *self-fertile* implies. Other trees rely on cross-pollination of blossoms from different tree varieties within the same species to produce a crop. Whether the tree is self-fertile or requires pollination, yields always increase when more than one tree variety is present. Not all fruiting trees blossom at the same time so it's important to consider pollinating partners, or pollinator groups. These are trees of the same species of a different variety that are all in bloom simultaneously, allowing for cross-pollination between trees. Pollinator group A sets their blossoms earliest in the spring, followed by pollinator group B, C, and eventually D toward the end of the spring season.

Size and Yield: Espaliered, Dwarf, Semi-dwarf, and Standard

As if chill hours and pollination groups weren't enough to take into consideration when selecting fruit trees, size also plays an important role. The height the tree reaches at maturity directly correlates to harvest, as a standard-sized tree has more fruiting wood available than a dwarf. Luckily, dwarf rootstocks require minimal space so even backyard growers with limited real estate can grow their own orchard fruit. Smaller trees reach maturity and are able to produce larger harvest quantities earlier than full-size tree varieties.

Espaliered: Not as much total yield but produces more fruit per square foot. Great for small-space growers, often grown in flat, two-dimensional shapes, and sets fruit in three to five years.

Dwarf: Roughly 8 feet (2.4 m) in diameter, produces full-sized fruit, but boasts a smaller yield. Dwarf stock can yield a harvestable crop in as little as three to five years.

Semi-Dwarf: These trees range from 12 to 15 feet (3.7 to 4.6 m) tall and wide, and can produce up to 500 fruits per season. They produce reliably for fifteen to twenty years after waiting an initial five years after planting for true harvest quantities.

Full/Standard Size: The largest of the fruiting trees, they produce anywhere from four to eight bushels per season or more (400 to 800+ pounds [181 to 363+ kg]) depending on species. At maturity, full- or standard-sized trees reach an average 20 to 30 (6 to 9 m) feet in height.

Harvesting cherries from a 100+-year-old tree native to the property.

THE CONTRIBUTION OF FRUIT AND NUT TREES TO THE HOMESTEAD

The following is a list of common fruiting trees and the contributions with which they can provide the homestead. Note all harvest amounts are an average, based on ideal growing conditions and after initial establishment of the tree.

SPECIES	AVERAGE YIELD / SEASON 1 BASKET = 1 BUSHEL	TIMING	TREE CONTRIBUTION
APPLE	Standard Semi-dwarf Dwarf	Chill hours: **200–1000** Harvest: **AUTUMN**	Fruit harvest for fresh eating, preserving, wine, cider, jam, firewood, lumber for furniture and cabinetry, blossoms for pollinators, medicinal properties, apple cider vinegar, wood chips
APRICOT	Standard Semi-dwarf Dwarf	Chill hours: **500–800** Harvest: **SUMMER**	Fruit harvest for fresh eating and jams, oil used in cosmetics and skincare products, medicinal properties, hardwood for decorative woodturning and musical instrument making, slow-burning firewood
CHERRY	Standard Semi-dwarf Dwarf	Chill hours: **800–1200** Harvest: **SUMMER**	Fruit harvest for fresh eating and preservation, food source for birds and wildlife, woodworking, woodturning, wood chips, flooring, food smoking, firewood, medicinal properties
ALMOND	Standard 50+ pounds (23+ kg) Semi-dwarf 20-35 pounds (9–16 kg) Dwarf 20 pounds (9 kg)	Chill hours: **400–600** Harvest: **AUTUMN**	Fresh eating, flour-making, almond milk, oil for cosmetics and skincare products, Ayurvedic medicine, woodturning, slow-burning firewood, blossoms for pollinators
PEACH	Standard Semi-dwarf Dwarf	Chill hours: **200–800** Harvest: **SUMMER**	Fruit harvest for fresh eating and preservation, jams, wine, woodworking, clean-burning wood stove fuel, blossoms for pollinators
NECTARINE	Standard Semi-dwarf Dwarf	Chill hours: **100–500** Harvest: **SUMMER**	Fruit harvest for fresh eating and preservation, blossoms for pollinators
PEAR	Standard Semi-dwarf Dwarf	Chill hours: **400–900** Harvest: **AUTUMN**	Fruit harvest for fresh eating and preservation, woodworking, musical instrument production, furniture, cabinetry, wood chips, blossoms for pollinators, medicinal properties

SPECIES	AVERAGE YIELD / SEASON 1 BASKET = 1 BUSHEL	TIMING	TREE CONTRIBUTION
PLUM	Standard Semi-dwarf Dwarf	Chill hours: **400–700** Harvest: **SUMMER**	Fruit harvest for fresh eating and preservation, plum brandy, plum kernel oil, fine woodworking, woodturning
POMEGRANATE	Standard 150+ fruits Cold hardy 100 fruits Semi-dwarf 75+ fruits	Chill hours: **100–200** Harvest: **AUTUMN**	Fruit harvest for the seeds used in fresh eating and baking, juice, medicine, dye-making, high tannins used for curing leather, bark used as insecticide in some countries
MULBERRY	Standard 15-25 pounds (7–11 kg) Semi-dwarf 10+ pounds (5+ kg) Dwarf 5+ pounds	Chill hours: **200–400** Harvest: **SUMMER**	Fruit harvest for fresh eating, jams, juice, wine, trap crop to protect cherry trees, food source for birds and wildlife, dye-making
QUINCE	Standard Semi-dwarf Dwarf	Chill hours: **200–500** Harvest: **AUTUMN**	Fruit harvest for jams, jellies, marmalades, puddings, wines, medicinal properties, fine-textured wood coveted for fine woodworking, woodturning, and artistic furniture inlays
MEDLAR	Standard 25+ pounds (11+ kg) Semi-dwarf 10+ pounds (5+ kg) Dwarf not available	Chill hours: **200–400** Harvest: **AUTUMN**	Fruit harvested for candy making, jams, jellies, pickling, marmalade and syrup, creates medlar cheese when combined with eggs and butter, baked, roasted, cider, wine, brandy, medicinal properties, high tannin levels make bark and leaves good for tanning hides, flexible wood good for fishing-rod production
PAWPAW	Standard not available Semi-dwarf 25–50+ pounds (11–23 kg) Dwarf 5+ pounds (2+ kg)	Chill hours: **400+** Harvest: **AUTUMN**	Harvestable fruit for fresh eating or cooking, crushed leaves used as insect repellent, dried pliable bark can be twisted into rope, tinder, fire kindling, wood-carving substrate
FIG	Standard 50+ fruits Semi-dwarf 20+ fruits Dwarf 10+ fruits	Chill hours: **100–200** Harvest: **SUMMER**	Harvestable fruit for fresh eating, cooking, dehydrating, preservation in jams and jellies, medicinal properties
PERSIMMON	Standard Semi-dwarf Dwarf	Chill hours: **200–400** Harvest: **AUTUMN**	Harvestable fruit for fresh eating, cooking, baking, jams; strong wood used for golf clubs, drumsticks, pool cues, and tool handles; blossoms for honey bees and native pollinators
PECAN	Standard 200+ pecans Semi-dwarf 40+ pecans Dwarf 10+ pecans	Chill hours: **300–500** Harvest: **AUTUMN**	Harvestable nuts for eating raw, baking, source of food for wildlife; strong wood for flooring, furniture, meat smoking; was used in automobile framing and agricultural machinery; leaves are an excellent mulch for their slow decomposition rate

Pruning

A healthy fruiting tree is better equipped to resist pests and disease naturally. Installing guilds certainly helps with attracting predatory insects, and even deterring some forms of rust and blight, but a pruned tree is structurally sound in the first place. Pruning at the right time of year (when the tree is dormant during colder months) does not hurt the tree, nor does it pose a threat to its overall health and productivity. In fact, the opposite is the case. A tree free of diseased and decayed wood is better able to produce fruit, as its energy is no longer focused on health and abscission.

Trees should be pruned when dormant so open wounds can scab and heal before the free flow of airborne bacteria and pests. My rule of thumb is to remove the Three Ds—dead, decayed, and diseased material. This is done by cutting the branch or shoot at an angle parallel to the base of the main branch or tree trunk. The cut should be made at the crotch, or intersection, of the shoot being removed from the main branch. A saw or sharp scissors are required so as not to cause tearing, which opens up the tree to disease.

Excessive growth is also removed to open up the canopy and allow for adequate airflow. Sunlight is better able to reach the fruit and help with ripening. The act of pruning improves the harvestable yield of the fruit by encouraging the health of new and healthy fruiting wood. It also allows for the grower to better control the size of the tree. If a more squat or bush-like tree with many lateral side shoots is desired, simply cut off the whip, or centermost branch of the tree, during planting. This will encourage the tree to create side shoots.

A member of the sheep flock snacks on a low-hanging branch of the black walnut tree.

ORCHARD TREE PRUNING SHAPES

Different tree species may require different pruning shapes. A peach tree, for example, is commonly pruned into an open-vase as it's said to increase fruiting. Espaliered trees are often grown against walls or other flat surfaces, so two-dimensional pruning may be required. Below is an illustration of the most common pruning structures.

ESPALIERED
Custom shaping to accommodate a wall or to be trained on a support structure.

A. Horizontal Cordon
B. Candelabra
C. Fan
D. Belgian Lattice

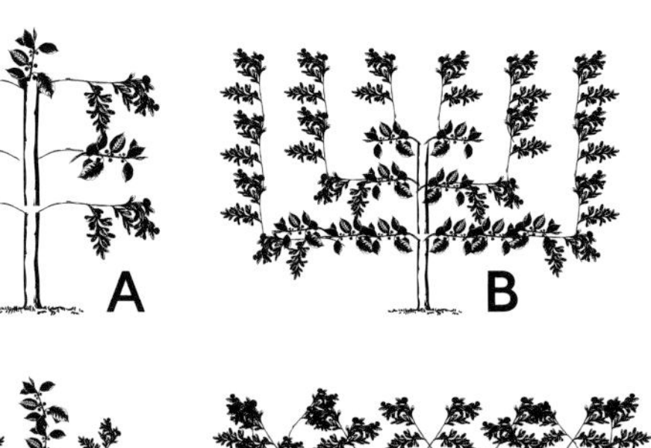

Suitable for:
Dwarf stock
Fruit and nuts

CENTRAL LEADER
One main central stem growing vertically with several side branches, evenly spaced.

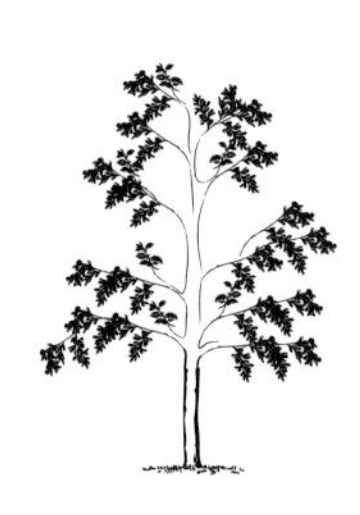

Suitable for:
Dwarf stock
Semi-dwarf
Standard
Fruit and nuts

MODIFIED CENTRAL LEADER
A modified leader has the top portion of the leader removed with several layers of strong lateral branches.

Suitable for:
Dwarf stock
Semi-dwarf
Standard
Fruit and nuts

OPEN CENTER
The leader is completely removed while the tree is still young and 3 to 5 secondary branches are trained to grow open like a bowl. Maximizes fruit production while managing the overall height of the tree.

Suitable for:
Semi-dwarf
Standard
Peach
Nectarine
Plum

CREATING AN UNDERSTORY: PLANTING IN GUILDS

I believe one of the best visuals of permaculture planting systems is that of an orchard tree guild. A guild is a micro-ecosystem of plants that work together to support the main fruiting tree, creating layers of harvestable, functional vegetation. This layered approach to companion planting also provides the grower with more food in a given space by growing upwards rather than thinking in terms of growing outwards.

When I learned about guilds and their function, I already had my young fruit trees in place. I had installed apples, pears, cherry trees, peaches, plums, and a cold-hardy pomegranate. Because my trees are sprinkled intermittently throughout the farm, I chose to create circular and square shaped guilds immediately surrounding the base of my trees. Linear guilds may be a better fit for trees that reside in rows.

I started by collecting large cardboard boxes, free of heavy inks and varnishes, and removed any tape and adhesives. I flattened them and placed them directly on top of the grass and any weeds that were growing at the trunk's base. Over time, cardboard would decay resulting in added organic matter for the soil. But first, the smothered weeds and growth beneath the cardboard would wither and die back, also returning matter and nutrients to the soil. Next, I piled compost directly on top of the cardboard, several inches thick. The weight of the compost keeps the cardboard in place and a new growing space is created. I researched the best guild members for each variety of fruiting tree I was growing. Then I began planting guild members directly into the freshly laid compost.

The Seven Members of a Guild System

There are seven members in an orchard tree guild. The first participant is the tree, such as an apple or pear tree. Immediately surrounding the centerpiece, within the mature drip line of the canopy, are supporting characters each of which provides a service. There are nitrogen fixers that pull nitrogen from the atmosphere and fix it into the soil's layers. They provide a readily available source of nitrogen to the main tree. Another role is that of insect repellers. Crops such as calendula, lavender, and nasturtium repel unwanted insects that can cause damage and ultimately disease. Simultaneously, many of these same plants acts as pollinator attractors, inviting beneficial insects to the area to increase the amount of pollination for the fruiting tree. As a result, fruit production is increased.

In nature, no tree is found with a clean slate of mulch surrounding the trunk. Even if vegetation is sparse, there are always at least a few seedlings and groundcovers sprouting beneath. In a guild system, mulchers grow in between and around vegetation protecting the soil from the solarization of nutrients, erosion, and by withholding moisture. Weed suppressors also reside on the ground's surface, filling in gaps where weeds may choose to take hold. And finally, accumulators are plants with long taproot systems that have the ability to mine nutrients from deep within the soil and help to draw it upward for the benefit of the main fruiting tree. Many guild members can serve more than one function. For example, mint repels unwanted insects and is a food source for many beneficial pollinators and predatory insects.

ORCHARD GUILD MEMBERS BY TREE SPECIES

Species	APPLE
Pollinator Attractor	• Lavender • Nasturtium • Yarrow
Insect/ Disease Repellent	• Chives (deter apple scab) • Lavender (deters coddling moth) • Nasturtium (repels wooly aphids and squash bugs) • Tansy (deters Japanese beetles and squash bugs, toxic to humans)
Nitrogen Fixer	• Clover • Comfrey • Peas • Lentils • Beans • Chickpeas • Peanuts
Nutrient Accumulator	• Chicory (for potassium and phosphorus) • Tansy (mines potassium, toxic to humans) • Garlic (mines sulfur) • Yarrow • Valerian • Chickweed • Stinging Nettle • Sorrel • Vetch
Natural Mulcher	• Comfrey (trap crop for slugs) • Nasturtium • Alyssum • Rhubarb
Weed Suppressor	• Clover • Mint • Strawberry

Species	PEAR
Pollinator Attractor	• Bee Balm • Borage • Woodruff • Cosmos • Calendula • Yarrow
Insect/ Disease Repellent	• Chives • Marigold • Foxglove (improves disease resistance in pear trees) • Garlic • Nasturtium • Calendula
Nitrogen Fixer	• Red and White Clover • Beans • Alfalfa • Lupine • Peas
Nutrient Accumulator	• Chicory • Borage • Comfrey • Alfalfa • Garlic (mines sulfur) • Yarrow • Chickweed • Stinging Nettle • Sorrel • Vetch
Natural Mulcher	• Comfrey • Hostas • Buckwheat • Nasturtium • Alyssum • Rhubarb
Weed Suppressor	• Mint • Strawberry • Woodruff

Species	PEACH
Pollinator Attractor	• Cosmos • Sunflowers • White and Red Clover (attract Trichogramma wasps) • Yarrow
Insect/ Disease Repellent	• Chives • Garlic (prevents peach curl and vine borers) • Rosemary (for slugs and snails) • Lavender (for slugs and snails) • Thyme (for slugs and snails)
Nitrogen Fixer	• Comfrey • Clover • Peas • Lentils • Beans • Chickpeas • Peanuts
Nutrient Accumulator	• Chicory (for potassium and phosphorus) • Garlic (mines sulfur) • Yarrow • Chickweed • Stinging Nettle • Sorrel • Vetch
Natural Mulcher	• Comfrey • Nasturtium • Alyssum • Rhubarb
Weed Suppressor	• Mint • Strawberry • Clover

Species	PLUM
Pollinator Attractor	• Cosmos • Yarrow • Daylilies
Insect/ Disease Repellent	• Garlic (repels aphids and flea beetles) • Lemon Balm • Chives
Nitrogen Fixer	• Comfrey • Clover • Peas • Lentils • Beans • Chickpeas • Peanuts
Nutrient Accumulator	• Chicory • Garlic (mines sulfur) • Yarrow • Chickweed • Stinging Nettle • Sorrel • Vetch
Natural Mulcher	• Comfrey • Nasturtium • Alyssum • Rhubarb
Weed Suppressor	• Mint • Clover

In nature, no tree is found growing in isolation. Guild companions help to repel unwanted insects and even disease, attract pollinators, mine nutrients from the soil, fix nitrogen, suppress weeds, and act as mulch. Members of a guild vary based on tree species.

ORCHARD GUILD MEMBERS BY TREE SPECIES

Species	CHERRY	
Pollinator Attractor	• Calendula • Chamomile (attracts predatory insects of oriental fruit moth, tarnished plant bug, and cherry fruit fly) • Coriander • Daisy	• Alyssum • Yarrow • Dill • Fennel • Buckwheat • Butterfly Weed • Mustard Greens
Insect/ Disease Repellent	• Chives • Calendula, Chamomile, and Oregano all act as antifungals • Lemongrass • Lemon Balm	• Marigold • Nasturtium • Garlic • Onion • Mulberry (acts as trap crop for birds)
Nitrogen Fixer	• Comfrey • Clover • Peas • Lentils	• Beans • Chickpeas • Peanuts
Nutrient Accumulator	• Lupine (potassium and phosphorous) • Garlic (mines sulfur) • Yarrow • Valerian	• Chickweed • Stinging Nettle • Sorrel • Vetch
Natural Mulcher	• Comfrey • Nasturtium	• Alyssum • Rhubarb
Weed Suppressor	• Clover • Strawberry • Vining Winter Squash • Mint	• Buckwheat • Pennyroyal • Thyme

Species	POMEGRANATE
Pollinator Attractor	• Lavender • Daisies • Yarrow
Insect/ Disease Repellent	• Nasturtium • Fennel
Nitrogen Fixer	• Comfrey • Clover • Peas • Lentils • Beans • Chickpeas • Peanuts
Nutrient Accumulator	• Chicory • Yarrow • Chickweed • Stinging Nettle • Sorrel • Vetch
Natural Mulcher	• Comfrey • Nasturtium • Alyssum • Rhubarb
Weed Suppressor	• Mint

Species	QUINCE
Pollinator Attractor	• Yarrow • Sage • Borage • Chives • Daylilies • Comfrey • Chicory • Dill • Fennel • Coriander
Insect/ Disease Repellent	• Garlic • Egyptian Walking Onions • Chives
Nitrogen Fixer	• Comfrey • Clover • Beans • Peas
Nutrient Accumulator	• Yarrow • Chicory • Comfrey • Horseradish • Chickweed • Stinging Nettle • Sorrel • Vetch
Natural Mulcher	• Comfrey • Nasturtium • Alyssum • Rhubarb
Weed Suppressor	• Strawberry • Wild Ginger

ORCHARD GUILD MEMBERS BY TREE SPECIES

Species	APRICOT
Pollinator Attractor	• Basil • Lavender • Yarrow • Daylilies
Insect/ Disease Repellent	• Garlic (for borers) • Wormwood (deters fruit flies) • Nasturtium (for aphids)
Nitrogen Fixer	• Clover • Comfrey • Peas • Lentils • Beans • Chickpeas • Peanuts
Nutrient Accumulator	• Lovage • Marjoram • Caraway • Garlic (mines sulfur) • Yarrow • Chickweed • Stinging Nettle • Sorrel • Vetch
Natural Mulcher	• Comfrey • Nasturtium • Alyssum • Rhubarb
Weed Suppressor	• Clover • Mint • Strawberry

Species	ALMOND
Pollinator Attractor	• Cosmos • Sunflowers • White and Red Clover (attract Trichogramma wasps) • Yarrow
Insect/ Disease Repellent	• Chives • Garlic (vine borers) • Rosemary (for slugs and snails) • Lavender (for slugs and snails) • Thyme (for slugs and snails)
Nitrogen Fixer	• Comfrey • Clover • Peas • Lentils • Beans • Chickpeas • Peanuts
Nutrient Accumulator	• Chicory (for potassium and phosphorus) • Garlic (mines sulfur) • Yarrow • Chickweed • Stinging Nettle • Sorrel • Vetch
Natural Mulcher	• Comfrey • Nasturtium • Alyssum • Rhubarb
Weed Suppressor	• Mint • Strawberry • Clover

Species	**PECAN** (note contains juglone, which may affect surrounding growth)
Pollinator Attractor	• Elderberry • Lemon Balm • Purple Coneflower • Bee Balm • Texas Blue Bonnet
Insect/ Disease Repellent	• Garlic • Chives
Nitrogen Fixer	• Comfrey • Russian Olive
Nutrient Accumulator	• Chicory
Natural Mulcher	• Comfrey
Weed Suppressor	• Strawberry • Lemon Balm • Black Raspberry

Guilds can be installed for more than just fruiting trees. This polyculture approach to planting is beneficial for most vegetation, including fruiting shrubs. Here at my farm, I surround my blueberry bushes with comfrey for fixing nitrogen and acting as a chop-and-drop mulch. Yarrow is an attractor of pollinators while mining nutrients from the soil. Tansy also is an accumulator, especially of potassium, simultaneously repelling Japanese beetles and squash bugs. Valerian pulls nutrients upwards for my blueberry bushes as well. I've added strawberry plants for weed suppression along with nasturtium as an insect repellent. Lettuce or spinach grow well among this guild for an additional edible crop. Rhododendrons and azaleas thrive in acidic soils so make excellent companions for blueberries if flowering shrubs are desired.

My raspberry plants are accompanied by garlic, tansy, rue, turnips, yarrow, and wormwood and are grown among pine trees. The pine needles drop and contribute acid to the soil, benefiting the raspberry canes, while acting as a natural mulch. The garlic is a pest deterrent warding off aphids, flea beetles, Japanese beetles, and spider mites. Tansy repels many insects including harlequin bugs, mines potassium, and attracts predatory wasps and beneficial pollinators. Rue is toxic to humans; however, it repels Japanese beetles. Turnips deter harlequin or stink bugs in my raspberry guilds, and wormwood is a great insect repellent too!

CHAPTER eight / THE ROLE OF THE HOMESTEADER

Cover crops. Companion planting. No till. Building and regenerating soil. Rotational grazing. Multispecies integration. These are only individual efforts. And while each piece is greatly important, when practiced together, the whole is greater than the sum of its parts: creating resilience in the land through sustainable solutions that can feed the soil that in turn feeds the people.

If our planet is healthy, we are healthy. This is permaculture. And just as nature is ever-evolving, so is the role of the homesteader. A homesteader must learn and be willing to pivot and to adapt to changes in the environment, weather, land, and animals. There is no blueprint for what the future holds nor for the appropriate responses to any ecological challenges. But Mother Nature is the ultimate teacher and will always offer guidance through demonstration. Through the observation of natural growing systems, wildlife interactions, and weather patterns, clues and answers lie.

« Cotton Patch geese are considered an endangered species by The Livestock Conservancy. We are currently breeding to help repopulate their numbers.

CONTINUE TO WORK WITH WILDLIFE AS CHALLENGES ARISE

Landscapes and farmscapes that are equipped with plant and animal diversity contain thriving soil structure, are host to perennials, and have the ability to retain moisture. They are more resilient to drought, severe weather events, erosion, soil solarization, water runoff, and pests or disease. As an advocate for the permaculture ecosystem, the homesteader has the responsibility of avoiding recurring problems by seeking natural solutions and asking for help from more seasoned farmers when needed.

For example, ticks are prevalent on my property. I happen to live in the most tick-infested county in the entire country (at the time this book was written). After I contracted Lyme disease from a tick bite, two of my dogs suffered from Lyme symptoms as well. When my animals were diagnosed with Lyme, I knew it was time to make a change. I introduced guinea fowl to the farm for their voracious appetites for ticks. They offer eggs I can either hatch out or sell. Their feet don't scratch growing spaces as they forage, and they also make excellent alarms—especially when they forage in areas that my geese don't frequent. Rather than fight the tick population with chemical solutions, I found a natural approach that offers stacked functions and fills a role no other animal can on my farm, and it will continue to work for me for years to come. To find this answer I had to rely on the wisdom of other farmers; those who had experience with guinea fowl as a means of tick control.

When the avian flu of 2022 began to largely affect both small farms and commercial poultry keepers, I needed to find a way to keep my birds from accessing our natural stream. Though it is rare to see wild birds populating our portion of the waterway thanks to our Livestock Guardian Dogs, I'm well aware that wild geese, ducks, and heron could be prevalent upstream. Migrating birds are the largest carrier of the virus and, as the homesteader, it became my job to minimize (if not completely eliminate) any potential contact or contamination of my flock. Very quickly we enclosed one of our pasture spaces with woven wire, equipped it with automatic waterers and kiddie pools.

The geese and ducks were housed within the pasture, far from the banks of the stream. As our guardians patrolled the exterior of the pasture, they were kept safe from intruding foxes and coyotes while landlocked. As a result of the constant inundation of duck and goose droppings on the pasture, forgeable growth exploded. Suddenly my pasture was lusher and greener than I can ever remember. My horses and sheep continued to rotate through the space, despite the new permanent tenants, and the land was happier for it. As grasses quickly reached roughly 3 feet (1 m) in height, the ducks began nesting within the tall growth providing cleaner eggs. They felt more safe and secure hiding within the vegetation just as their wild counterparts did. The avian flu initially was a challenge for the farm, but we adapted, and we were better for it.

With an increase in avian flu casualties across the northeastern United States, the flock is moved away from open waterways and contained in a pasture. This is an example of adaptation on the part of the homesteader.

Two Magpie ducklings just hatched and are bred on the homestead in an effort to assist in repopulating this species.

Right alongside my human efforts to mimic Mother Nature, she continued to adapt as well. I observed the introduction of the invasive and destructive spotted lanternfly into my state and watched as small nymphs quickly covered trees, fence rails, and even rose bushes. Suddenly praying mantis sacs could be found in abundance throughout my growing spaces. A food source was available, and the mantis population responded. I have since learned that these predatory insects are the number one consumer of the spotted lanternfly. By providing a permanent, sustainable home for them through permaculture practices, our farm is now host to the very answer we need to keep spotted lanternfly destruction in check. Barn swallows arrived shortly thereafter and could be seen swooping haphazardly through the air, feeding on the lantern flies as they dove, along with an increased numbers of bluebirds. An ecosystem is an ever-evolving balance of plant and animal life. The ecologically minded homesteader must accept this natural progression and embrace the ongoing strategic thinking and problem-solving that accompanies it.

REPOPULATE THREATENED HERITAGE SPECIES

Heritage breeds of livestock and poultry are those that were brought to the United States generations ago and employed for multiple purposes on the farm. A goose, for example, was expected to provide eggs, meat, excellent foraging and weeding capabilities, and act as a guardian. There was no room on the primitive homestead for animal species that performed only one function or merely looked attractive. As time went on, however, many heritage goose breeds were forgotten, and birds with flashy feathers were sought after for their prizewinning fluff. Specific traits were sought rather than a well-balanced offering of many—a goose was wanted purely for their ability to lay a large number of eggs per season, or their flavorful meat only. As a result, many common heritage breeds of geese and other animals have declined.

As a permaculture homesteader, I am interested in these heritage species for their multipurpose services. Geese that can help to weed my pastures, forage well, require little supplemental feed during seasons of growth, have friendly dispositions, are watchful guardians, and provide eggs and fertilizer as they graze are, in my opinion, excellent candidates when creating a working ecosystem. I started incorporating Large Dewlap Toulouse, Sebastapol, and Cotton Patch geese into my flock and, just recently, I am happy to say we have started a breeding program. We are working to help rebuild their numbers as they assist in building our sustainable homestead. I also keep Ancona, Cayuga, and Magpie ducks, Clydesdale horses, and Shetland sheep; all have a place on the endangered or threatened list (at the time this book is being written). By providing these species with a home, I am capitalizing on the stacked functions of each species and contributing to building the greater whole that is their population.

PLANT FOR FUTURE GENERATIONS

Often growers think of the harvest they'll reap when sowing seeds and planting crops. After all, receiving a yield is likely the main intention when growing food. But incorporating perennial crops requires time before a plentiful harvest is available. Sure, fruit and nut trees will bear fruit in small amounts when they are young. But it can take years, maybe even a decade or more, before the tree will boast a full offering.

In a society accustomed to instant gratification, it's so important to remember that incorporating perennials benefits not only the current farmer, but the farmers for generations to come. The fruits of your labor will provide years of food, wood, carbon absorption, and shelter for wildlife and insects, as well as facilitate soil health and structure. In one year, a mature tree will absorb more than 48 pounds (22 kg) of carbon dioxide from the atmosphere, releasing oxygen in exchange. It will sequester roughly 1 ton (1,000 kg) of carbon dioxide in the soil by the time it reaches forty years old. A mature fruiting tree can produce 150 to 300 pounds (68 to 136 kg) of harvestable fruit, depending on the species, variety, and adequate pollination. Fertilizer trees contribute to soil health by absorbing nitrogen from the atmosphere and pulling it into the soil through the roots and leaf litter. They extract and bring up nutrients from deep within the soil for the benefit of other crops whose roots reside within more shallow layers of the rhizosphere. The roots also help to stabilize erosion, increase water infiltration, improve soil density, and reduce water runoff.

The Contribution of Deciduous Fruiting Trees
It is much easier to track the amount of produce a plant will provide over the course of one growing season, than to measure a yield over the course of the plant's lifetime. I find it fascinating to learn how much food, carbon absorption and nutrients one tree can contribute to an ecosystem. Assuming an apple tree is well pruned, pollinated, and cared for with minimal pest and disease damage, the average standard-sized apple tree living an average of forty years will yield 8 to 10 bushels of apples per season (variety dependent). Dwarf and semi-dwarf varieties will yield one to two bushels per year, approximately. While fruit size varies, on average one bushel contains 126 apples. That's the potential to reach 1,260 apples per season, and 37,800 apples during the course of one tree's thirty-year productivity time. (It takes eight to ten years for an apple tree to become truly productive). That is an enormous amount of fruit for one household. One study also found that just 1 acre (0.4 ha) of apple trees pulls about 20 tons (18 metric tons) of carbon dioxide from the air each season, releases 15 tons (14 metric tons) of oxygen, and provides over five billion BTUs of cooling power. Some carbon is sequestered in the new woody tissue and root systems as well.

As for other common fruiting orchard trees, standard-sized pear trees produce reliably after about four years. Dwarf and semi-dwarf varieties mature in about two or three. Under ideal conditions, standard trees may yield 4 to 6 bushels, or 1 to 2 bushels for dwarfs and semi-dwarfs. There are approximately 48 to 50 pears per bushel. At an approximate lifetime of fifty years, minus the first few for tree establishment, that's 8,832 pears minimum. Similarly, standard-sized peach trees offer 4 to 6 bushels per season and 1 to 2 bushels per year for dwarf rootstocks. A study conducted in South Korea found that a twenty-five-year-old cherry tree has the ability to absorb 20 pounds (9 kg) of carbon emissions annually, thus acting as a major contributor to greenhouse gas reduction.

The Contribution of Hard Wood Trees
Many of the hardwood trees that offer quality materials for saw or lumber mills also provide a harvestable crop for use on the homestead. Maple trees provide one of the hardest wood materials for woodworking while their sap is prized for maple syrup making. Oak trees and their giant canopies provide decades of shade to the understory, shelter for wildlife. Acorns also can be a source of food for livestock and humans. (Note acorns are only safe for human consumption after tannins have been successfully removed.) Chestnut and walnut species provide nuts for cooking, roasting, consuming raw, or for sharing with the local wildlife population. In addition to their nut yields and timber, these massive trees create an overstory in a permaculture growing system. They serve as windbreaks and sun catchers while absorbing and sequestering massive amounts of carbon both in their woody tissues and by pulling it into the soil. The gigantic, well developed root systems run deep, loosening compacted layers, assisting in the prevention of erosion, and drawing up soluble nitrogen from deep within the soil.

All trees have the ability to pull carbon dioxide from the atmosphere and use it in the development of plant tissue, roots, leaves, bark, and buds. But those that live longer, grow taller and have large trunk diameters absorb more carbon over the course of their life-span. By the time a sugar maple tree reaches twenty-five years old, it will have sequestered roughly 400 pounds (181 kg) of carbon dioxide. Researchers have also calculated that a silver maple traps approximately 25,000 pounds (11 metric tons) of carbon dioxide after fifty-five years. This is twenty-five times more carbon than cherry

While black- and raspberries prefer full sun, they grow quite well within forest growing spaces at Axe & Root Homestead. They receive speckled sunlight in the forest understory.

and plum trees. Oak trees can absorb about 92 pounds (42 kg) of carbon per year while the rapidly growing American chestnut tree has been measured to intake 60.1 metric tons of carbon per hectare for a nineteen-year-old tree.

The Contribution of Conifers

Pine, spruce, fir, hemlock, and a few other softwood conifer trees make up roughly 80 percent of the paper industry, and their trunks are sought after for logging and milling. Bark and shavings are used in animal bedding; pine sap is usable as an ingredient in turpentine and resins; and tar is created by the high temperature carbonization of pine wood in anoxic conditions. When young, fir tips, pine tips, and spruce tips are edible and contain large quantities of vitamin C. As the needles age, pine, spruce, and fir needles dry and can be used as mulches for growing spaces and as fire starters, and they are said to possess medicinal properties. Cedar wood is highly resistant to rotting and is used for decking, trim work, housing siding, and fencing. The small, blue juniper berries are edible when properly prepared and when eaten in moderation. They also are the ingredient that gives gin its unique flavor.

In the living pine tree forests in east Georgia, up to 220 Gg of carbon dioxide is sequestered annually. One mature cedar tree can intake approximately 48 pounds (21.77 kg) of carbon dioxide each year. And as with any tree, the roots act as a living filtration system, catching rainfall and filtering it of pollutants and contaminants, all while absorbing it from the soil.

The Power of Perennial Fruit and Vegetable Crops

While fruit, nut, and hardwood trees are the heavy hitters when it comes to carbon absorption and sequestration, planting perennial vegetable crops absolutely should be included in any ecologically minded homestead portfolio. One plant can return for decades, providing harvestable yields each season. These shrub and understory plants take on the role of suppressing weeds and keeping the soil shaded. Some act as living mulches, and all contribute to healthy soil structure by contributing nitrogen and other nutrients. Deeper root systems are more drought tolerant and are able to withstand flooding events.

Asparagus beds can be productive under the right growing conditions for up to thirty years. Each season, an average harvest per plant is roughly a ½ pound (226 g) of spears. A healthy artichoke should live for up to six years and can provide up to ten buds or more per season. Chives, if encouraged and well cared for, will come back year after year, providing edible greenery each spring. These fragrant blades also are excellent companions for both the garden and the orchard as their blossoms attract beneficial insects and their strong odor repels pests.

As for fruiting crops, an individual raspberry plant can live for up to ten years if well maintained. And as the canes continue to send out runners, new plants are always being established. This means a single raspberry plant can produce offspring, supplying a food source for decades. In addition, these perennial berry bearers provide a shelter and food source for wildlife and birds. Rhubarb returns every spring with a life expectancy of seven to ten years. It is common for most rhubarb plants to provide 2 to 6 pounds (907 g to 2.7 kg) of food per growing season.

Easy Action Items to Implement Now as a Sustainable Homesteader

- Incorporate more perennials into your diet and landscapes.
- Facilitate the soil-animal nutrient cycle by allowing rotational grazing, crop rotation, and animal-plant integration when possible.
- Employ rotational grazing to encourage more photosynthesis in pasture spaces, thus intaking more carbon.
- Reduce waste and recycle when possible.
- Feed soil first by adding to soil structure. Feed plants second by way of healthy soil.
- Include more heirloom crop varieties in growing spaces and in diet. Hybrids and F1 varieties are bred for faster yields, pest resistance, and sweeter taste, which depletes soil structure and health. Microbes and mycorrhizae are unable to keep up with rapid plant growth rates.

THE KITCHEN IS AN EXTENSION OF THE HOMESTEAD

Most of the plants that humans eat largely focus on just one portion of the crop. For example, growing an entire corn stalk for the ears. Harvesting just the heads of broccoli and cauliflower. Harvesting snap peas for their inner peas and tossing the tough or stringy pods. There are some instances where plant leaves, stalks, or stems may be toxic, and those areas of the plant should not be ingested. In many other instances, however, the remaining plant material can be consumed by livestock or, at the very least, be added to a compost heap. In some cases, the crop can even be regrown if given the right conditions. By using as much of the plant as possible, food waste is eliminated, more food per plant is consumed, and animals benefit from additional forage and nutrients.

Homegrown Food Scraps for the Animals

Every Halloween, pumpkins and gourds are carved into jack-o'-lanterns. But all the pulp, seeds, and innards are discarded. Share the leftover pulp and seeds (orange pumpkins and gourds only) with horses, cows, sheep, pigs, and goats. They'll receive a boost of vitamin C, and the natural compound within called *cucurbitin* helps to ward off parasites. Raw and cooked pumpkin flesh and seeds are safe for dogs to consume. Cats can have pumpkin as well so long as it's fully cooked. Chickens and ducks also love to snack on a freshly cracked open pumpkin.

The table on page 177 provides a general guide to sharing common garden or produce waste material with livestock. For the health and safety of animals, never feed moldy or spoiled material and always be sure to discuss feeding scraps to your livestock with your veterinarian before doing so. None of the items listed are a replacement for a balanced diet.

Food Scraps for the Garden

Keeping food scraps out of landfills reduces harmful greenhouse gas emissions into the atmosphere. Revisiting the pumpkin example, there are ways to benefit from pumpkin and gourd scraps before adding them to the compost, throwing them away, or sharing them with livestock. Consider roasting the seeds for a nutritious snack. You can also separate and discard the strings, and briefly rinse the seeds. Allow to dry on a towel and save the seeds for next year's garden. The seeds from tomatoes, peppers, eggplant, summer squash, cucumber, and more are easy to save by removing the individual seeds from the pulp. Tomato pips benefit from soaking or rinsing to remove stubborn pulp before drying.

Lettuce heads contain both the leaves and the base where the crop was initially cut from the stem. If you remove the leaves for consumption, the base of the lettuce head can be saved and placed in a shallow dish of water. It will regrow an entirely new head of lettuce if placed on a windowsill or replanted in the garden and given access to sunlight. The same goes for the base of fennel, bok choy, scallions, leeks, and celery. A potato that is wrinkled can still be used as a seed potato for the garden. Simply plant it whole and don't separate into individual sections.

When carrots and beets are harvested from the garden, their green tops are typically removed from the root base when consuming. Save the top of the carrot or beet where the root connects to the green stems. If placed in a shallow dish of water or outdoors in the garden, carrot greens or beet greens will regrow (the root itself will not). These greens can be used in sauces, smoothies, and more. Sometimes sweet potatoes begin to wither after long storage periods. Just plant an entire sweet potato tuber in a pot or in the garden if the weather permits;

Kitchen Scraps for Livestock	
Lettuce	Feed leaves to horses, cows, sheep, goats, pigs, geese, ducks, and chickens.
Watermelon	Finely chopped rind, seeds, and flesh are safe for cows, goats, pigs, ducks, geese, and chickens. The flesh is considered safe for sheep.
Strawberry	Strawberries and hulls are safe for pigs, ducks, chickens, geese, cows, and goats. Avoid feeding to sheep and horses.
Kale	Leaves and stems can be fed in small quantities to cows, goats, sheep, pigs, and geese. Never horses; cruciferous vegetables produce gases that may lead to colic.
Radish and Turnip	Greens may be fed to cows, sheep, goats, pigs, chickens, ducks, and geese so long as no sprouts or blossoms are included. Flowers contain mustard oil, which can be toxic to animals. Horses can be offered greens in moderation.
Corn	Corn husks chopped into small pieces (to avoid choking hazards) can be fed to pigs, cattle, goats, and sheep. Stalks are very fibrous and are considered safe forage for cows and pigs. Sheep can eat stalks with slow acclimation. Goats can consume stalks in moderation.
Peas	Pea pods, both fresh and dried, are considered safe for goats and pigs. Finely chopped or ground peas and pods are preferred for cattle. Chopped peas and pods are safe for horses, sheep, ducks, chickens, and geese.
Beets	Greens are a favorite of geese, ducks, chickens, cows, and horses. Greens contain oxalic acid; this can bind with calcium to cause a deficiency in goats and sheep, so feed in moderation.
Apples	Horses, pigs, and cows love apples, and they are considered safe in moderation. Very small pieces may be fed to goats and sheep infrequently. Small chunks are favored by geese, chickens, and ducks.

new shoots will sprout from the eyes of the sweet potato. Carefully snip the shoots at the base and transfer to a glass of water. Roots will begin to develop and can be transplanted to a garden container or outdoor growing space to generate an entirely new crop of sweet potatoes.

Bean and pea pods can be difficult to find on the vines. If you come across a dried yellow or brown pod that is hardened and crisp, harvest the beans or peas from within and save these to plant for the next gardening season. After the cap has been removed, mushroom stems can be inserted into moist, warm soil. New mushrooms of the same variety should sprout. (Be sure to have experience when harvesting mushrooms before consumption to avoid ingestion of dangerous or toxic look-alikes.)

Citrus seeds may sprout if tossed into the compost pile and can also potentially increase the acidity. Instead, immediately place the seeds from oranges, lemons, limes, and grapefruit into a dish of warm water to soak. Keep them hydrated until you are able to plant in a small container with seed starting mix. With adequate water and sunlight, a new citrus tree will be born. Corn stalks and sunflower stems are both woody and upright. After harvesting the heads of the flowers and the ears of corn, cut the tough stems with a saw or pair of branch trimmers at the soil line. Bring the long stems into a garage or basement to dry and cure throughout the winter months. Come spring, these woody poles can be reused and repurposed for creating trellises in the garden.

Preserve the Fruits of Labor

When the land is healthy, the vegetation is healthy, and thriving plants yield abundant harvests for the homesteader. Ideally, the grower will have enough homegrown produce to harvest and eat fresh, to preserve, and to share with both barnyard family members and the local community. The kitchen is most certainly an extension of the holistic homestead as homemade goods require less transport than store-bought items, reduce carbon emissions, provide more nutrients because they are picked at ideal harvesting times when nutrients are at their peak, reduce food waste, and eliminate unnecessary packaging.

Freezing

As soon as fruit and vegetables are harvested, chemical compounds called enzymes begin the process of deteriorating the color, texture, and taste of the produce. For this reason, it's imperative to blanch vegetables in boiling water for one minute before freezing. The act of boiling deactivates the enzymes and also has the benefit of reducing the amount of living microorganisms on the surface of the vegetable that can lead to spoilage. After a brief boiling period, the produce must be plunged into icy water to stop the cooking process. This prevents the softening of the product as well as a loss of nutrients.

Fruit, because it often is served raw, is not blanched like vegetables. Coat the picked fruit with ascorbic acid (vitamin C) and stir to ensure it's evenly coated before freezing. The vitamin C will prevent any fruit browning. Regardless of blanching or using ascorbic acid, produce can be spread into a single layer on a baking sheet and frozen for twenty minutes. Remove the tray from the freezer and package the produce in its permanent freezer-safe container. The act of flash freezing this way before packaging helps to prevent clumping and an abundance of ice buildup within the package. When properly prepared and placed within an airtight freezer-safe container, frozen produce can retain almost all of its nutritional content—more than many other food preservation approaches, in fact. During the height of the growing season, the amount of produce being harvested can be overwhelming. Freezing is a relatively quick option for preserving fresh vegetables and fruit. Items such as tomatoes that may be intended for canning as sauce, paste, and salsa later on, can be frozen whole in the interim. When more time allows, the tomatoes can be thawed and canned.

Freeze-Drying

With the recent introduction of home freeze-drying machines, growers, homesteaders, and culinary experts are turning to this method of food preservation. Lyophilization or freeze-drying results in a food product retaining much of its original texture, flavor, nutritional content, and, if prepared properly, a shelf life of roughly ten years or more based on conservative estimates. The finished item is lightweight, easy to store, and incredibly effective for the preservation of foods containing low acidity, eggs, and milk products.

During the freeze-drying process, foods are frozen and placed under a vacuum. Then sublimation occurs (the scientific principle of the direct transition of a solid to a gas). In the instance of freeze-drying, ice from the frozen foods is evaporated, rather than first converting to liquid (water). Because of this transformation from ice to evaporation, foods retain its flavor, texture, and nutritional content once rehydrated.

Canning

All canning methods are not equal when it comes to preserving homegrown produce. Water-bath canning tends to be a more approachable, beginner canning regimen. It is the act of processing acidic canned goods housed in glass canning jars, affixed with metal bands and lids, in a large pot of boiling water. Processing times vary based on the item being canned. Peaches are a common water-bath canned item as they contain high levels of acid that aid in shelf stability. Tomato sauce, paste, salsa, and whole or stewed tomatoes also are good candidates for water-bath canning when combined with a small amount of lemon juice to achieve the right acidity. Jams and jellies also are commonly water-bath canned. The avoidance of botulism and food spoilage is best achieved by following scientifically tested and proven water-bath canning recipes from credible sources. Water-bath canned home goods should be used within one year of production. Most vitamin and mineral content can be maintained with the exception of vitamins A and C. A loss of 5 to 20 percent of these vitamins has been reported per year stored.

Pressure canning is the right approach for any low-acid foods, meat-based items, pickles, broths, prepared goods such as soups and stews, or generally any homemade recipe you wish to can. Pressure canning achieves the temperatures of 240°F (116°C) or higher by way of steam and pounds of pressure. These high temperatures are required within the jar to stabilize these food items for storing, while water-bath canning cannot. Nutrient loss is roughly the same as in water-bath canning and should also be used within one year of the date of preparation.

Pickling

A balanced combination of acid, spices, sugar, and produce creates a pickled vegetable. It's essential to follow scientifically proven recipes from reputable sources in order to achieve crisp, flavorful products that are considered safe for ingestion. Cucumbers, beans, carrots, radishes, asparagus, beets, and turnips are all commonly pickled. Items should be clean and free of bruising and decay. Some harvested produce requires blanching before packing into jars so be sure to follow recipe guidelines closely. Yeast and mold spores are the two major contributors to spoilage in pickled foods. Some pickling recipes call for heat processing in a water bath or pressure canner before consumption to reduce these threats. Refrigerator pickling recipes require no heat and typically last only about one week before expiration.

Lacto-Fermenting

Fermenting foods is relatively safe and easy so long as well-researched and proven recipes are followed. During the lacto-fermentation process, *Lactobacillus*, a bacteria already present on the surface of the fruit or vegetable, feeds on the sugars within, converting them into lactic acid. Lactic acid is a natural preservative. This beneficial bacteria can survive salt used within fermentation brine recipes while harmful bacteria cannot. Research shows this ages-old method of food preservation also increases the nutritional content of some vegetables and even adds B12, a vitamin not naturally occurring in plants, while boosting their overall shelf life and flavor.

A diet rich in fermented foods contains a myriad of health benefits. Probiotics, the naturally occurring beneficial bacteria in fermented items such as pickles, kimchi, and sauerkraut, contribute to digestive health, a reduction in inflammation within the body, and overall health of the gut microbiome.

Dehydrating

Drying or dehydrating food material—with the right balance of airflow, humidity, and a bit of warmth—removes the majority of water from within the fruit or vegetable. This process also retains almost all of the food's nutritional value, prevents mold and bacteria from growing, and creates a shelf stable good. Ovens, sun-drying, air-drying, and dehydrators all facilitate the dehydration process and humidity levels, temperature, and drying times vary.

Once an item is fully dried, it needs to be stored in an airtight container safe from air flow and insects. Check the item after a couple of days of packaging to see if any moisture has collected within the container. If condensation is present, remove the item, and check for mold or mildew. If these are present, discard the item and sterilize the package for use another time. If the partially dehydrated items are still good and present no signs or odors indicating spoilage, further dry the material and be sure it's fully dehydrated and allowed to cool before packaging in a new container. Dehydrated fruit and vegetables take up very little space, weigh little because water weight has been evaporated, and retain much of their original flavors.

Share the Abundance

Knowledge is power. And so is food. When the planet is happy, the people are happy. These concepts shape one of the three major tenants of permaculture: Fair Share or, simply put, share the surplus with the people and return it to the land. Here at the farm, I choose to donate approximately one-third of my harvests in the peak summer season to the local food pantry. Fresh picked flowers are offered at the farm stand for free. I try to share my experiences and the insights I've learned through teaching via social media platforms and at public speaking events. To me, it seems that the more people who become acquainted with the benefits of working directly with the land will, in turn, become inspired and wish to facilitate their own homegrown operation and care for Mother Nature too.

I believe we live in a society largely focused on the individual and on independence from one another. But it's not possible to know everything. It's not possible to grow everything, to keep every animal, and to make or produce every homegrown good imaginable. There simply aren't enough hours within a day. This is where bartering, help and assistance, sharing, and supporting local farmers and growers comes into play. Sustainably minded homesteaders and growers can create networks that benefit everyone, and most definitely the land, too. Imagine entire communities striving to create land resilience and locally sourced food while fostering solutions that work in alignment with nature. I believe it all begins with one sustainable homestead.

Water bath canning is often used on the homestead for preserving food products with a high acid content. For low-acid goods, pressure canning is required.

Resources

Recommended Reading

Brown, Gabe. *Dirt to Soil: One Family's Journey into Regenerative Agriculture.* Chelsea Green Publishing, 2018.

Phillips, Michael. *The Holistic Orchard: Tree Fruits and Berries the Biological Way.* Chelsea Green Publishing, 2012.

Shepard, Mark. *Restoration Agriculture: Real-world Permaculture for Farmers.* Acres USA, 2013.

Online Tools

USDA Plant Hardiness Zone Map

Schonbeck, Mark, and Ron Morse. "Choosing the best cover crops for your organic no-till vegetable system." Rodale Institute. Rodale Institute 29 (2004).

Stock Density Calculator, Millborn Seeds

Grazing Period Stock Density, University of Maine–Cooperative Extension

Conservation Priority List, The Livestock Conservancy

Helpful Products

Luster Leaf 1625 Digital Soil Thermometer

Luster Leaf 1601 Rapitest Test Kit for Soil pH, Nitrogen, Phosphorous, and Potash https://www.amazon.com/Luster-Leaf-1601-Rapitest-Phosphorous/dp/B0000DI845

HIRALIY Compost Tea Air Bubbler

Spalding Laboratories Fly Predators

Ball Water Bath Canner

Harvest Right, Home Freeze Dryer

Maple Tapper Store, Deluxe Maple Syrup Tree Tapping Kit

Acknowledgments

This book was created ultimately as a result of my children. Being a mother led me down a path of living that I didn't know existed before them. Thank you to my two sons, my husband, my friends, and my family for supporting me and encouraging me while researching, writing, and designing the infographics for this book. You all ran the marathon with me, cheered me on toward the finish line, and I'm so grateful and thankful to have you all. You know who you are.

And thank you to my editor, Thom O'Hearn, who truly heard my idea for this, helped me to shape into something usable, and worked tirelessly to bring this book to life.

About the Author

Angela Ferraro-Fanning is a self-taught, first-generation farmer who built Axe & Root Homestead, a 6-acre (2.4 hectares) farm in central New Jersey. Originally a graphic and website designer by trade, she owned and operated her own design firm for over a decade. After the birth of her first child, she realized she wanted to be outdoors, aligning her life with the seasons and with nature. She now grows and preserves her own homegrown produce for her young family and runs a farm bustling with Clydesdales, geese and ducks for eggs, an apiary with ten beehives, sheep, and a small orchard. She is the author of The Little Homesteader series of books as well as *The Harvest Table Cookbook*. She currently shares her love for eco-conscious, self-sufficient living with others through social media as @axeandroothomestead, her online homesteading classes, public speaking, and regular contributions to publications such as *Countryside*, *Backyard Beekeeping*, and *Backyard Poultry*.

Works Cited

Chapter 2

Brown, Gabe. *Dirt to Soil: One Family's Journey into Regenerative Agriculture*. Chelsea Green Publishing, 2018.

Davis, Donald R., Melvin D. Epp, and Hugh D. Riordan. "Changes in USDA Food Composition Data for 43 Garden Crops, 1950 to 1999." *Journal of the American College of Nutrition* 23, no. 6 (2004): 669–682.

Institute for Agriculture and Trade Policy, GRAIN, Greenpeace International. "New Research Shows 50 Year Binge on Chemical Fertilisers Must End to Address the Climate Crisis." November 1, 2021.

Philpott, Tom. "New Research: Synthetic Nitrogen Destroys Soil Carbon, Undermines Soil Health." *Grist*, February 23 (2010).

SARE Outreach. *Keeping Fertilizer in the Ground and Out of the Air.* 2015/2016 Report from the Field, Western Sustainable Agriculture Research and Education.

Schonbeck, Mark, and Ron Morse. "Choosing the Best Cover Crops for Your Organic No-till Vegetable System." *Rodale Institute*, 29 (2004).

Singh, Balwant, and Darrell G. Schulz. "Soil Minerals and Plant Nutrition." *Nature Education Knowledge* 6, no. 1 (2015): 1.

Chapter 4

Compare Duck Breeds. Metzer Farms.

El-Hack, Abd, Mohamed T. El-Saadony, Ahmed R. Elbestawy, Ahmed R. Gado, Maha M. Nader, Ahmed M. Saad, Amira M. El-Tahan, Ayman E. Taha, Heba M. Salem, and K. A. El-Tarabily. "Hot Red Pepper Powder as a Safe Alternative to Antibiotics in Organic Poultry Feed: An Updated Review." *Poultry Science* 101, no. 4 (2022): Art-101684.

Grant, Amy. "Cover Crops Chickens Eat: Using Cover Crops for Chicken Feed." *Gardening Know How*, November 12, 2021.

Hutchins, Paul, and Betsy Hutchins. "Why Mules?" Rural Heritage.

The Livestock Conservancy Quick Reference Guide to Heritage Ducks. The Livestock Conservancy.

Tanner, Savannah. "Selecting Cattle for Forage Efficiency." University of Georgia Cooperative Extension, June 8, 2020.

Chapter 5

Barnes, Amber. "Things That Are Toxic to Geese." *The Open Sanctuary Project*, September 13, 2021.

College of Agriculture, Food and Environment. "Multispecies Grazing." University of Kentucky.

College of Agriculture, Food and Environment. "Using Cover Crops for Grazing Cattle." University of Kentucky.

Ehrhardt, Richard. "Cover Crop Grazing with Sheep: Lessons Learned from Recent MSU Extension Demonstration Event." Michigan State University Extension, December 15, 2011.

Forteau, L., Bertrand Dumont, Guillaume Salle, G. Bigot, and Géraldine Fleurance. "Horses Grazing with Cattle Have Reduced Strongyle Egg Count Due to the Dilution Effect and Increased Reliance on Macrocyclic Lactones in Mixed Farms." *Animal* 14, no. 5 (2020): 1076–1082.

Goat Pastures Poisonous Plants Photodynamic. Extension Foundation and the USDA Cooperative Extension. August 14, 2019.

Grant, Amy. "Cover Crops Chickens Eat: Using Cover Crops for Chicken Feed." *Gardening Know How*, November 12, 2021.

Grazing Period Stock Density, University of Maine–Cooperative Extension

Hart, Steve. "Parasite Control with Multispecies and Rotational Grazing." Langston University, 2014.

Hartman, David. "Cover Crops for Livestock Grazing." PennState Extension, July 3, 2014.
Huntington, Peter. "Alternative Hays for Horses." *Equinews Nutrition & Health Daily*, May 22, 2012.

How Do I Manage Internal Parasites in Multispecies Grazing? *ATTRA – Sustainable Agriculture Program.*

Kentucky Equine Research Staff. "Use of Cover Crops in Horse Pastures." *Equinews Nutrition & Health Daily*, September 13, 2021.

Kephart, Kenneth, Gilbert R. Hollis, and D. Murray Danielson. "Forages for Swine." Originally published as PIH-126. *Swine*, Extension Foundation and the USDA Cooperative Extension, August 28, 2019.

Knight, Carole. "More Than a Tin Can—Forage Systems for Goats." University of Georgia Cooperative Extension, June 1, 2018.

PennState Extension. "What Kinds of Forages Can We Use for Swine?" *Swine Production and Management Home Study Course.*

Spalding Laboratories. "Fly Predators Biology." February 7, 2022.

Stock Density Calculator, Millborn Seeds.

Strickler, Dale. "Pasture Management for Horses." *Green Cover* January 15, 2018
USDA Department of Agriculture. "Cover Crop Technical Note no. 16: Cover Crops Used by Wildlife, Bees, and Beneficial Insects." Revised July 6, 2021.

Wheaton, Howell, and John Rea. "Forages for Swine." University of Missouri, Department of Animal Sciences.

Young, Amy. "Salmonellosis" UC-Davis, Veterinary Medicine, Center for Equine Health, August 28, 2020.

Chapter 6

Apps, Rozie. "How to Make Hot Compost in 4 Weeks." Permaculture, November 1, 2013.

Compost Use and Soil Fertility. University of Massachusetts Extension, January 2013.

Compost. University of California, Division of Agriculture and Natural Resources.

Composting At Home. United States Environmental Protection Agency.

Nutrient Value of Composts. University of California Cooperative Extension.

Schwarz, Mary, and Jean Bonhotal. "Composting at Home-The Green and Brown Alternative." Cornell Waste Management Institute, 2011.

Want to Keep Your Compost Weed-Free? Weed Science Society of America, April 6, 2009.

Chapter 7

Blumenstock, Marvin (Bud), and Kathy Hopkins. "How to Tap Maple Trees and Make Maple Syrup." University of Maine–Cooperative Extension.

Bramen, Lisa. "Chugging Maple Sap." *Smithsonian Magazine*, March 10, 2011.

Dalibard, Christophe. "Overall View on the Tradition of Tapping Palm Trees and Prospects for Animal Production." *Livestock Research for Rural Development* 11, no. 1 (1999): 1–37.

Engels, Jonathon. "Black Locust, How Do I Love Thee, Let Me Count the Ways: Hmm. At Least Seven." Permaculture Research Institute, June 16, 2017.

Farmer, Sarah. "Black Locust & Drought." USDA Southern Research Station, April 2, 2020.

Feeley, Chris. "Black Walnut: The Killer Tree." Iowa State University.

Gedefaw, Mohammed. "Environmental and Medicinal Value Analysis of Moringa (*Moringa oleifera*) Tree Species in Sanja, North Gondar, Ethiopia." *American International Journal of Contemporary Scientific Research* 2, no. 9 (2015): 20–36.

Goat Vegetation Oklahoma Browse. Extension Foundation and the USDA Cooperative Extension, August 14, 2019.

Morbeck, George C. "Pecan: Its Chief Characteristics and Its Many Uses." (1927).

Sang-Hun, Choe. "The Forests of Southern Korea Yield a Prized Elixir." *The New York Times*, February 24, 2009.

Stafne, Eric. "Chilling-Hour Requirements of Fruit Crops." Mississippi State University Extension.

Chapter 8

Boeckmann, Catherine. "Planting Apple Trees and Harvesting Apples." Almanac.

Botzek-Linn, Deb, and Suzanne Driessen. "Pickling Basics." University of Minnesota Extension.

Could Global CO_2 Levels be Reduced by Planting Trees? CO2Meter, May 03, 2021.

Driessen, Suzanne, and LouAnn Jopp. “Drying Food at Home.” University of Minnesota Extension.

Grazing Corn Stalks with Beef Cattle. PennState Extension, October 29, 2018.

Harvard Health Publishing. “Fermented Foods Can Add Depth to Your Diet.” Harvard Medical School, April 19, 2021.

Jacobs, Douglass F., Marcus F. Selig, and Larry R. Severeid. “Aboveground Carbon Biomass of Plantation-grown American Chestnut (Castanea dentata) in Absence of Blight.” *Forest Ecology and Management* 258, no. 3 (2009): 288–294.

Lakso, Alan L. “Estimating the Environmental Footprint of New York Apple Orchards.” *New York Fruit Quarterly*, 18 (2010): 26–8.

Making Paper from Trees. Forest Service, United States Department of Agriculture, Revised March 1997.

Marinelli, Janet. “Nature’s Champion Carbon Eaters.” National Wildlife Federation, March 15, 2011.

Merrill, Cathy, Brian Nummer, Christine Jessen, Paige Wray, and Callahan Ward. “Buying a Home Freeze-Dryer: What to Know Before You Go.” Utah State University, Preserve the Harvest Extension.

Morrical, Dan, and Joseph Rook. “Grazing Corn Stalks with the Ewe Flock.” Ohio State University, College of Food, Agriculture, and Environmental Sciences, November 3, 2020. (Previously published with MSU College of Veterinary Medicine—Sheep publication: December 10, 2008).

Nummer, Brian, and Brandon Jahner. “Storing Canned Goods.” Utah State University, Preserve the Harvest Extension.

Rust, Steven, and Dan Buskirk. “Feeding Apples or Apple Pomace in Cattle Diets.” *Cattle Call* 13, no. 4 (2008): 2–3.

Schafer, William, and Suzanne Driessen. “The Science of Freezing Foods.” University of Minnesota Extension, reviewed in 2021.

Shim, Elizabeth. “Cherry Trees Absorb Carbon Emissions, South Korea Scientists Say.” UPI, April 7, 2020.

Stancil, Joanna Mounce. “The Power of One Tree—The Very Air We Breathe.” *Forestry*. U.S. Forest Service, June 03, 2019.

Trees Improve the Soil. Common Pastures, February 20, 2016.

Vachnadze, G. S., Z. T. Tiginashvili, G. V. Tsereteli, B. N. Aptsiauri, and Q. G. Nishnianidze. “Carbon Stock Sequestered from the Atmosphere by Coniferous Forests of Eastern Georgia in Conditions of Global Warming.” *Annals of Agrarian Science*, 14, no. 2 (2016): 127–132.

Wadhwa, Manju, M. P. S. Bakshi, and H. P. S. Makkar. “Utilization of Empty Pea (Pisum sativum) Pods as Livestock Feed.” *Broadening Horizones*, 46 (2017): 1–4.

Index